THERMO-MECHANICAL MODELING OF ADDITIVE MANUFACTURING

THERMO-MECHANICAL MODELING OF ADDITIVE MANUFACTURING

Edited by

MICHAEL GOUGE
PAN MICHALERIS

Butterworth-Heinemann
An imprint of Elsevier

Butterworth-Heinemann is an imprint of Elsevier
The Boulevard, Langford Lane, Kidlington, Oxford OX5 1GB, United Kingdom
50 Hampshire Street, 5th Floor, Cambridge, MA 02139, United States

Library of Congress Cataloging-in-Publication Data
A catalog record for this book is available from the Library of Congress

British Library Cataloguing-in-Publication Data
A catalogue record for this book is available from the British Library

ISBN: 978-0-12-811820-7

For information on all Butterworth-Heinemann publications
visit our website at https://www.elsevier.com/books-and-journals

Working together
to grow libraries in
developing countries

www.elsevier.com • www.bookaid.org

Publisher: Matthew Deans
Acquisition Editor: Christina Gifford
Editorial Project Manager: Anna Valutkevich
Production Project Manager: Nicky Carter
Designer: Mark Rogers

Typeset by VTeX

Contents

I

THE FUNDAMENTALS OF ADDITIVE MANUFACTURING MODELING

1. An Introduction to Additive Manufacturing Processes and Their Modeling Challenges

MICHAEL GOUGE AND PAN MICHALERIS

2. The Finite Element Method for the Thermo-Mechanical Modeling of Additive Manufacturing Processes

MICHAEL GOUGE, PAN MICHALERIS, ERIK DENLINGER, AND JEFF IRWIN

II

THERMOMECHANICAL MODELING OF DIRECT ENERGY DEPOSITION PROCESSES

3. Convection Boundary Losses During Laser Cladding

MICHAEL GOUGE

4. Conduction Losses due to Part Fixturing During Laser Cladding

MICHAEL GOUGE

5. Microstructure and Mechanical Properties of AM Builds

ALLISON M. BEESE

11. Mitigation of Distortion in Large Additive Manufacturing Parts

ERIK R. DENLINGER

III

THERMOMECHANICAL MODELING OF POWDER BED PROCESSES

12. Development and Numerical Verification of a Dynamic Adaptive Mesh Coarsening Strategy for Simulating Laser Power Bed Fusion Processes

ERIK R. DENLINGER

13. Thermomechanical Model Development and In-Situ Experimental Validation of the Laser Powder-Bed Fusion Process

ERIK R. DENLINGER

14. Study of the Evolution of Distortion During the Powder Bed Fusion Build Process Using a Combined Experimental and Modeling Approach

ALEXANDER J. DUNBAR

List of Contributors

Allison M. Beese
The Pennsylvania State University, USA

Lei Chen
Mississippi State University, USA

Long-Qing Chen
The Pennsylvania State University, USA

Erik R. Denlinger
Product Development Group, Autodesk Inc., State College, PA, United States

Alexander J. Dunbar
Applied Research Laboratory, The Pennsylvania State University, State College, PA, USA

Michael Gouge
Product Development Group, Autodesk Inc., State College, PA, United States

Jarred C. Heigel
The Pennsylvania State University, University Park, PA, USA

Jeff Irwin
Product Development Group, Autodesk Inc., State College, PA, United States

Yanzhou Ji
The Pennsylvania State University, USA

Pan Michaleris
Product Development Group, Autodesk Inc., State College, PA, United States

About the Editors

Pan Michaleris is a Senior Software Architect at Autodesk Inc. He received his Ph.D. in 1994 in theoretical and applied mechanics at the University of Illinois at Urbana-Champaign, and was a senior research engineer at the Edison Welding Institute (EWI) until 1997. Pan served as a professor in the Mechanical and Nuclear Engineering Department at the Pennsylvania State University from 1997 to 2016. In 2012, Michaleris founded and served as both president and lead developer at Pan Computing LLC. Pan Computing was a software development and commercialization company for physics-based modeling of additive manufacturing processes. Pan Computing was acquired by Autodesk Inc. in 2016. His areas of interest include computational mechanics, finite element methods, manufacturing process modeling, and residual stress and distortion. Michaleris authored *Minimization of Welding Distortion and Buckling* and in addition to more than 80 peer reviewed journal and proceedings papers. He formerly served on the editorial board of *Science and Technology in Welding and Joining*, and was an associate editor for *Welding Journal*.

Michael Gouge is a Research Engineer for Autodesk where he focuses on validating and improving the thermo-mechanical modeling of additive manufacturing processes. He completed a Ph.D. in mechanical engineering at the Pennsylvania State University under Dr. Pan Michaleris. His graduate research was on the finite element modeling of heat transfer, distortion, microstructure, and material properties of directed energy deposition processes, an area in which he has authored several peer reviewed journal articles. He has also received a B.A. in philosophy from Austin Peay, a B.S. in mechanical engineering from the University of Texas San Antonio, and a M.S. in mechanical engineering from Penn State.

Acknowledgments

CHAPTER 1

The editors would like to thank Christina Gifford and Anna Valutkevich for making this book possible.

CHAPTER 2

The authors would like to thank Pan Michaleris for developing and documenting the modeling work contained in this chapter.

CHAPTER 3

The material is based upon work supported by the Office of Naval Research through the Naval Sea Systems Command under Contract No. N00024-02-D-6604, Delivery order No. 0611. Jarred C. Heigel is supported by the National Science Foundation under Grant No. DGE1255832. Any opinions, findings, and conclusions or recommendations expressed in this material are those of the authors and do not necessarily reflect the views of the National Science Foundation. The authors would like to thank Pan Computing LLC for the generous use of their computing resources and access to the Netfabb Simulation FE code. The authors would also like to thank Guy Showers and Douglas E. Wolfe for providing the surface roughness measurements.

CHAPTER 4

The material is based upon work supported by the Office of Naval Research through the Naval Sea Systems Command under Contract No. N00024-02-D-6604, Delivery order No. 0611. The authors would like to thank Pan Computing LLC for the generous use of their computing resources and access to the Netfabb Simulation FE code.

CHAPTER 6

Yanzhou Ji and Long-Qing Chen acknowledge the financial support from the American Makes National Additive Manufacturing Innovation Institute (NAMII) under grant number FA8650-12-2-7230. Lei Chen is grateful for the financial support by the Start-up funding and the cross-college working group grant from Mississippi State University. The authors acknowledge Dr. Fan Zhang at CompuTherm LLC for providing the Ti-Al-V thermodynamic database; Dr. Alphonse A. Antonysamy at GKN Aerospace for providing the microstructure figures of additively manufactured Ti-6Al-4V. The authors are also grateful for Dr. Tae Wook Heo at Lawrence Livermore National Laboratory and Dr. Nan Wang at McGill University for useful discussions on model development.

CHAPTER 7

The authors gratefully acknowledge the financial support of America Makes - Multi-Sensor Thermal Imaging for Additive Manufacturing. This work is built upon early efforts under ONR SBIR. The authors would like to acknowledge ARL Penn State and Penn State University's Center for Innovative Material Processing through Direct Digital Deposition (CIMP-3D) for use of facilities and equipment. This material is based on research sponsored by Air Force Research Laboratory under agreement number FA8650-12-2-7230. The U.S. Government is authorized to reproduce and distribute reprints for Governmental purposes notwithstanding any copyright notation thereon. The views and conclusions contained herein are those of the authors and should not be interpreted as necessarily representing the official policies or endorsements.

CHAPTER 8

This work was partially completed under funding from NSF Grant No. DGE1255832. Any opinions, findings, and conclusions or recommendations expressed in this material are those of the authors and do not necessarily reflect the views of the National Science Foundation.

CHAPTER 9

This work was sponsored by a subcontract from Sciaky Inc. and funded by AFRL SBIR #FA8650-11-C-5165. Jarred Heigel contributed into this work under funding from NSF Grant No. DGE1255832. Any opinions, findings, and conclusions or recommendations expressed in this material are those of the authors and do not necessarily reflect the views of the National Science Foundation. The authors would also like to acknowledge partial support of this research by the Open Manufacturing program of the Defense Advanced Research Projects Agency and the Office of Naval Research through Grant N00014-12-1-0840. Any opinions, findings, and conclusions or recommendations expressed in this material are those of the authors and do not necessarily reflect the views of the National Science Foundation, the Department of Defense or the U.S. Government. Distribution Statement A: Approved for Public Release, Distribution Unlimited.

CHAPTER 10

This work was sponsored by a subcontract from Sciaky Inc. and funded by AFRL SBIR #FA8650-11-C-5165.

CHAPTER 11

This work was sponsored by a subcontract from Sciaky Inc. and funded by AFRL SBIR #FA8650-11-C-5165. The authors would also like to acknowledge partial support of this research by the Open Manufacturing program of the Defense Advanced Research Projects Agency and the Office of Naval Research through Grant N00014-12-1-0840. Any opinions, findings, and conclusions or recommendations expressed in this material are those of the authors and do not necessarily reflect the views of the National Science Foundation, the Department of Defense or the U.S. Government. Distribution Statement A: Approved for Public Release, Distribution Unlimited.

CHAPTER 12

The author would like to thank Pan Computing LLC and Autodesk Inc. for the use of the Netfabb Simulation FE software.

CHAPTER 13

The author would like to thank Pan Computing LLC and Autodesk Inc. for the use of the Netfabb Simulation FE software.

CHAPTER 14

Laboratory activities conducted during this research were conducted at the Center for Innovative Materials Processing through Direct Digital Deposition at Penn State. This material is based on research sponsored by Air Force Research Laboratory under agreement number FA8650-12-2-7230 and by the Commonwealth of Pennsylvania, acting through the Department of Community and Economic Development, under Contract Number C000053981. The U.S. Government is authorized to reproduce and distribute reprints for Governmental purposes notwithstanding any copyright notation thereon. Any opinions, views, findings, recommendations, and conclusions contained herein are those of the author(s) and should not be interpreted as necessarily representing the official policies or endorsements, either expressed or implied, of the Air Force Research Laboratory, the U.S. Government, the Commonwealth of Pennsylvania, Carnegie Mellon University, or Lehigh University. The authors would like to thank Schlumberger-Doll Research for their support of this effort. Any opinions, findings, conclusions, and/or recommendations in this chapter are those of the authors and do not necessarily reflect the views of Schlumberger-Doll Research or its employees.

CHAPTER 15

This chapter was completed using work developed by the America Makes project "Development of Distortion Prediction and Compensation Methods for Metal Powder-Bed AM (Project 4026.001)." The authors would like to thank GEGRC, UTRC, and Honeywell Aerospace for their work in this project. The authors would also like to thank Pan Computing LLC and Autodesk Inc. for the use of the Netfabb Simulation FE software.

PART I

THE FUNDAMENTALS OF ADDITIVE MANUFACTURING MODELING

1

An Introduction to Additive Manufacturing Processes and Their Modeling Challenges

Michael Gouge, Pan Michaleris
Product Development Group, Autodesk Inc., State College, PA, United States

1.1 MOTIVATION

Additive manufacturing (AM) is experiencing a resurgence. While the public interest and knowledge of additive or more colloquially, 3D printing processes is a recent phenomena, this technology has be in use for four decades. Multipass welding, where repeated weldments are made not to join materials but to create freeform structures, is the first easily recognizable metallic additive process [1,2]. This was a popular topic of research from the 1970s through the late 1990s [3,4]. However the manual nature of such freeform builds did not readily lend itself to accurate, repeatable builds. In the 1980s the first direct energy deposition (DED) machines were built [5]. DED processes use a deposition material in either powder or wire form which is melted onto a substrate or built component using a laser or electron beam which is controlled numerically. DED builds are inherently more accurate and repeatable than multipass welding, which availed themselves to industrial practice. This became particularly popular after the first LENS™ machines were built by Sandia National Laboratories [6–8]. These offered near-net shape part builds at a cost which made them attractive alternatives to traditional manufacturing methods, notably casting. However the deeper industrial adaptation of AM processes did not come about until the creation and standardization of laser power bed fusion (LPBF) machines. While there are many variations, at its essence the LPBF process involves the spreading of a thin layer of powder on a surface which is melted by a quickly moving, very small heat source [9–11]. After each layer an elevator lowers the build plate by the thickness of the solidified powder layer, then a recoater blade pushes a fresh layer of powder over the built component, then the next layer is melted. This process is repeated until the part is completed. This methodology can create net shape parts with a much finer resolution than DED. Moreover, due to the design of the machines, the LPBF process is less onerous to setup for the additive engineer or technician, as instead of having to manually write machine code for DED controller, LPBF processes typically use a CAD source file which is separated into printable layers by software called a Slicer, which is then fed into the LPBF machine. This allows a high level of control over the builds with minimal effort or post-processing.

There is a great deal of excitement about AM technology. However, many new to the field find that the process is not as simple, accurate, or as repeatable as promised. This is due to the physical phenomena inherent to the process of melting a small amount of material upon a larger, cooler body. The expansion and contraction of both the added material and the existing component lead to permanent warping of the part and can leave high residual stresses [12,13]. This distortion imperils the economic feasibility of producing parts using AM technologies, as typically iterative builds are required to minimize part warpage until a useful part can be repeatably constructed [14–16]. Moreover, the material which is melted and then cooled experiences a much different temperature history than the base material which may be wrought or cast. The rate at which metals cool determine the particle size and metallurgic composition of the microstructure. This microstructure determines the mechanical properties of the solidified component, such as stiffness, elasticity, and yield strength [17]. These are unknowns to the design engineer who needs to produce a part within a certain tolerance with reliable material properties sufficient for the end use of said component. Many organizations resort to expensive iterative prints to develop printing parameters for each new geometry to produce an acceptable part. Each test build can take hours to days to complete and cost thousands to tens of thousand of dollars for each set of builds. Thus modeling of the thermo-mechanical and thermo-microstructure-property behavior is increasingly a useful or even necessary alternative to repeated experimentation.

Modeling AM processes is not without its own challenges. First one has to determine how to model the material addition. Then, there are inescapable non-linearities of the process, namely the temperature dependence of the thermal properties, the fourth order dependence of thermal radiation losses, plasticity, and large deformation modeling, which may need be accounted for. The boundary conditions, both thermal and mechanical, can be difficult to measure, model, approximate, and apply. The discretization of the problem in both space and time can require intense computational resources and a deeper understanding of computational methods to produce accurate models in a useful time frame. Finally, it is difficult to validate the modeling of AM processes, as the measuring process must be done in such a way as not to interfere with the normal operation of the machine, but also to give the insight necessary to ensure the model captures the behavior over the entire history of the process.

This book documents the details of modeling of AM processes with an emphasis on DED and LPBF systems. This chapter gives an overview of the processes themselves and the difficulties imposed to the prospective modeler. Chapter 2 outlines the numeric methods to describe AM and describes techniques to verify and validate the model. Chapters 3 and 4 rigorously describe methods applicable to convection and conduction losses, respectively, to improve the accuracy of an AM finite element (FE) model for laser cladding processes. Chapter 5 gives an overview of microstructure and material properties of common AM processes from an experimental metallurgist's view. Chapter 6 shows how phase-field modeling can be used to predict the evolution of microstructure during AM processes. Chapter 7 applies metallurgical principals to the thermal results of thermal FE models to predict microstructure and material properties of DED components. Chapter 8 is a larger overview of the thermo-mechanical modeling of thin walls built using the LENS^{TM} DED process. Chapters 9 and 10 concern the modeling of first small then large electron beam based DED while Chapter 11 illustrates how modeling of AM processes can be used to mitigate the distortion of large electron beam manufactured components. Chapter 12 shows the necessity of using advanced mesh coarsening methods for modeling LPBF processes and explains the techniques required to do this. Chapter 13 describes how to model LPBF builds using a moving source model and validates this process with in situ techniques. Chapter 14 introduces the multi-scale methods for prediction of distortion in powder bed processes, which is validated using in situ measurements, while the final chapter, Chapter 15, validates this modeling technique using post-process measurements.

1.2 ADDITIVE MANUFACTURING PROCESSES

Below is a brief overview of the common additive manufacturing processes. These AM processes all concern metallic systems, as the modeling of polymers is outside the purview of this book. The following list is not exhaustive, as there are many variants, research machines, and it is altogether likely new systems will emerge between the writing of this book and the reading of it.

1.2.1 Multipass Welding

While being the oldest form of additive processes and by practice the largest one, additive welding, such as metal inert gas welding, it is often left out of the discussion of AM processes. The argument against the inclusion of welding amongst the AM processes is that welding is most often described as a *joining* of two existent components, rather than the manufacture of a new component altogether. However this omission, based upon a very narrow definition of the term Additive Manufacturing, has the potential to limit the scope of research pursued by the additive community. The vast amount of both academic and industrial knowledge concerning various welding processes readily apply, in part or in toto, to both the modeling and practice of DED and LPBF manufacture.

The common additive welding processes – shielded metal arc welding (SMAW), gas metal arc welding (GMAW or MIG), wire fed tungsten inert gas welding (GTAW or TIG), flux cored arc welding (FCAW), and submerged arc welding (SAW) – are very similar from the prospective modeler's point of view. Each of these uses a wire feed for the additive material, which is melted by a plasma arc to join, fill, or repair parts, or even build new components using freeform techniques. These, as described, may sound identical to wire fed DED processes to the readers familiar with LENS™ etc. systems. However the principal difference is in the operation. DED processes are numerically controlled while welding is typically performed by a human operator.

Rosenthal laid the groundwork for the scientific analysis of welding, proposing an exponential, analytical solution for the heat distribution from a line heat source for welding [18]. Rosenthal's line heat source was used in much of the early numeric welding models. One of the first finite element welding models was created by Ueda, whose thermo-mechanical model was used to explore the elastic and plastic deformation caused by a single weld pass

[19]. This method was extended to multi-pass welding, showing that reasonable thermal and mechanical response results could be attained by combining weld passes into simultaneously modeled pairs [1]. Nickell and Hibbit produced and validated a more complete 2D welding model which included the effects of radiation, convection, and latent heat in the thermal model and accounted for the non-linearity of the mechanical properties [2]. Rybicki and Stonesifer used a similar non-linear 2D model to simulate multipass welding of pipe welds [3]. While thermal models compared favorably with experimental data, they attained less accurate results for residual stresses, particularly far from the weldment.

Improving the accuracy of thermal results necessarily improves the accuracy of the mechanical model, assuming that the mechanical model captures the elastic–plastic behavior of both the weld and joined materials. For the thermal analysis, more realistic boundary conditions were found to be necessary to achieve this end. Goldak et al. proposed the double ellipsoid heat source modeling method to replace Rosenthal's method [20]. Goldak showed that the double ellipsoid method produced thermal histories of single pass welds with much better agreement than Rosenthal's line input source. In the same paper the authors proposed an equation to represent the thermal losses from both convective and radiative losses at the melt pool surface.

Advances in the prediction of the mechanical response were enabled through first identifying the modes of distortion that occur during welding by Masubuchi [21]. Three of these distortion modes are in-plane and three are out-of-plane. The three in-plane modes are transverse shrinkage, longitudinal shrinkage, and rotational distortion. The three out-of-plane distortion modes are angular change, buckling, and longitudinal bending. From this basis, numeric models were improved by including viscoelastic effects [22], phase transformation [23], melt convection approximations [24–33] and moving from 2D to 3D FE analysis [31].

1.2.2 Directed Energy Deposition

Directed Energy Deposition uses a computer controlled moving heat source, typically a high powered laser or electron beam (EBEAM), to melt an additive material in either wire or powder form, which is fed into the melt pool. The primary difference between the two heat source types are the available power and the heat loss modes. Laser systems have a large range of possible powers, from around 100–5000 W [5,34]. Laser systems can be open to the atmosphere or within closed chambers, which is often filled with an inert gas to prevent contamination of the metal by atmospheric gases while in a molten state [35]. EBEAM systems can have much more throughput power, up to 10 kW [12]. The electron beam systems must be operated in a vacuum. This means that heat can only be pulled from the heated component through radiation and conduction. One final difference is the efficiency of these two heat sources. Only a small amount of the incident heat from a laser is actually absorbed by the built component, typically 30–50% [36,37]. EBEAM, in contrast, has an extremely high efficiency, typically 90–95% [12].

Many systems, such as the common LENS^{TM} machines, use a 3-axis motion controller which can move in the Cartesian directions [6,38]. These 3-axis processes can produce a wide variety of parts using a layer-by-layer deposition. Less common are the multiaxis systems, which add one or more rotational degrees of freedom up to a full 6 axis system. This can be accomplished through the rotation of the deposition head, the build platform, or a combination of the two. Multiaxis systems allow for a much greater range of manufacturable geometries. Material may be deposited in any direction using a full range of movement. While this design freedom is a boon to the AM production engineers, this necessarily increases the complexities for the simulation of such processes.

At its essence, additive manufacturing is an extension of additive welding technology: a moving heat source melts added material onto an existing substrate. Thus, the same analytic tools and scientific understanding of the earlier technology may be applied to the later one, with the caveat that AM processes introduce new complexities and new challenges. The scale of deposition requires that the deposited material be accounted for in the model. Furthermore, it increases the size of the models and thus the time required to compute them. Many AM parts cannot be modeled in 2D nor have a line of symmetry, which implies that the whole build must be incorporated into a 3D FE mesh. A higher number of deposition tracks necessitates a higher cumulative input power, which can translate, depending processing parameters, into much higher temperatures throughout the substrate. This has the effect of increasing the non-linearity of the problem, slowing down the already distended computational time. The increased non-linearity due to high temperatures is compounded by radiation losses, which has a quartic temperature dependence. Heightened surface temperatures also means that buoyancy effects, namely natural convection, will become non-negligible. Many direct deposition AM systems utilize inert gas flows to both propel powder into the melt pool and shield the melt pool from environmental contamination [36,39]. These gas streams create a localized region subject to forced convection [40–42]. Finally, the extended elevated temperature field frequently reaches the contacting regions of the substrate, where the build plate is clamped to a work surface or fixtured in a single or double cantilever system. Thus in some instances, to fully account for the distribution of heat, the fixturing body itself must be incorporated into the model [43].

Modeling of single lines of deposition using DED processes has been successfully modeled using 2D methods in the early days of DED research [5,44–47]. These models have been experimentally validated, but they are limited in both their scope, not trying to predict full thermal histories, distortion or residual stress, and also limited to planar-symmetric geometries. Thermal models using 3D finite elements expanded the range of geometries and processes that could be simulated using numeric methods [36,37,39,48–56]. Thermo-mechanical models, used to predict distortion or residual stresses during DED processes from the modeled thermal histories [12,53,55,57–60]. This is of utmost importance as distortion and stresses are the primary impediment to the wider adoption and implementation of AM into the manufacturing chain through the economic wastage of numerous failed parts. However there has also been extensive research documenting the significant differences in the microstructure and resulting mechanical properties of AM components [6–8,17,61–76]. Thermo-material models have sought to use the thermal model instead to predict microstructure evolution and the resulting material properties [50,77,78]. These models could ensure a priori that DED properties are sufficient for their end use, or even be used to alter processing conditions to design mechanical properties [79].

1.2.3 Laser Powder Bed Fusion systems

Laser Powder Bed Fusion components are built similarly to DED processes, using a numerically controlled heat source, building a part layer-by-layer, with the essential difference being that in LPBF machines powder is pre-placed upon the build plate instead of fed directly into the melt pool [10]. This powder is melted by the laser, in the areas specified by the machine controller, then the build platform is lowered, and a new layer of powder is spread over the build platform by a recoater blade. This allows for the layer-by-layer manufacture of parts. Other differences are that the effective heat source diameter in LPBF machines tends to be much smaller than DED, customarily around 1/10th to 1/20th the size, on the lower range of power, ranging usually between 100–250 W, and using a much faster deposition speed, 50–100 times common DED travel speeds, which results in LPBF heat source speeds between 500–1000 mm/s [54]. The minuscule size of the heat source allows for a much finer resolution than DED builds, creating fully net shaped parts. Yet the same difficulties of distortion, residual stress, and unknown material properties persist. Indeed there are two additional failure types to contend with, *Support structure failure* and *Recoater blade interference.*

Support structures are small volume sacrificial builds used to mitigate LPBF part distortion [16]. These structures are built in tandem with the desired component, but with thin walls, often in lattice structures, with limited, frequently sawtooth connections to the build plate and fully dense component. The purpose of these structures is to conduct heat away from the part during deposition and to hold the component down to the build plate during deposition. Post build, supports are cut away from the build plate and component, and are recycled or otherwise discarded. Support structure failure occurs when the immense stresses that are common during LPBF manufacture sever the less dense, small cross-section support from the build plate or component [80]. This failure allows the build component to distort to the point of not being usable, or may even lead to recoater blade interference, which will be described below.

The recoater blade is what sweeps the powder over the build chamber between laser melting passes. The recoater can be metallic, polymer, or even a roller. Recoater blade interference occurs when the upward distortion of the part is so great that it pushes above the next layer's powder, into the path of the powder sweeping device. When the recoater sweeps the next layer, it impacts the component. For a pliant polymer or roller system, this can decrease the effectiveness of the powder spreading process in subsequent layers. For metallic recoater blades this can be much graver, as the recoater blade can be bent or broken and the component itself damaged by the impact. This means that such a failure requires a new build, a new blade, and perhaps further machine repair.

Thus the modeling efforts to predict stresses and distortion of LPBF processes have extended uses, trying not only to protect the integrity of the build, but to preserve the functionality of the machines and safety of the operators. Below a brief summary of LPBF modeling efforts is described, focusing upon those works which have attempted to predict or mitigate part distortion.

The same methodologies of welding and DED modeling have been applied to the numeric simulation of LPBF processes. However, the relatively minor differences, particularly the size of the laser beam and the speed of deposition, create immense modeling demands. The mesh size depends upon the size of the laser and the length of the time steps is inversely proportional to the speed of the heat source travel speed. Thus a LPBF model will require a very dense mesh and extremely small time steps, which both contribute to excessively long simulation times.

The above limitations have motivated research focused upon capturing the behavior of single lines to small LPBF components [54,81–87] all of which used direct, moving source modeling methods using a fixed mesh. Similarly, small build thermo-mechanical models have been explored by Matsumoto et al. [81], Dong et al. [88], Dai and Gu [89]. These were all validated using post-process measurements. Other modeling attempts go so far as trying to model the powder particle effects [90–92]. For lattice like structures, beam theory based models investigate loading of lightweighted geometries [93–95]. To improve the speed and decrease the computational requirements of LPBF simulations, Zeng et al. produced a dynamically coarsening FE mesh algorithm which decreased the thermal simulation time five times [96]. Dunbar et al. used an analogous coarsening technique to perform a full thermo-mechanical simulation of a small LPBF build using measurements of both temperature and distortion for validation [97].

The above simulations show that direct simulations of LPBF processes are impractical with current computational systems due to the excessive time and memory requirements, which do not help the industrial practioner. To elide the computational requirements of doing moving heat source simulations, researchers have investigated analytical means [98], convolution methods [99] or layer-by-layer simulations [100,101]. This last methodology rests upon the assumption that due to the high rate of deposition, each layer can be considered a single time step, with an assumed or approximated deposition temperature and a simplified solution of the process mechanics to predict warping and stresses. However the current solutions in the literature are limited in their ability to capture geometry with varying thicknesses and scalability. An improved multi-scale simulation approach which has surmounted these limitations will be described and validated in Chapters 14 and 15.

1.3 CHALLENGES IN THE FINITE ELEMENT MODELING OF AM PROCESSES

With the usefulness of AM processes and the importance of creating accurate models of such processes now established, the challenges of these simulations can now be briefly outlined. Despite the differences in the AM technologies laid out above, modeling all of these methods must surmount the following challenges:

1. Modeling the addition of material
2. Modeling the heat input
3. Accurately accounting for thermal losses during building
4. Accurately modeling elasto-plastic stresses and strains
5. Coupling the thermo-mechanical behavior in a manageable way
6. Accounting for the temperature dependence of material properties from room temperature through melting

These are the basic requirements for any model of welding, DED, EBEAM, or powder bed deposition. While the above list is already daunting, there are further challenges presented by these processes which may have to be resolved to perfect a model, including:

1. Changes in microstructure
2. Changes in material properties
3. Anisotropy in thermo-mechanical material properties
4. Phase transformation effects

These two lists of concerns form the primary challenges to overcome in order to model AM processes in their totality. However, there is a greater challenge still that is tacitly linked to all of the above issues to overcome: completing simulations in a useful time frame. If it takes weeks or months to complete a thermo-mechanical simulation it would be wiser to accept the costs of iterative experimental prints to complete the build within the time demanded by the customer. A simulation tool must give useful results in a small fraction of the build time, so that multiple simulations may be completed if necessary to mitigate distortion and residual stresses before the first part is printed. Below, each of the listed challenges will be briefly described. The remainder of this book details the degree to which each item presents a challenge and shows experimentally validated techniques for overcoming each modeling challenge.

1.3.1 Material Addition

The first challenge in the modeling of Additive Manufacturing processes in explicitly stated in the umbrella term for these technologies, the addition of new material into the model. At its essence, the addition of material into a numeric model is the addition of new equations. There are two competing methods to account for new equations, referred to commonly as the *Quiet* method and the *Dead-Alive* method [50,56]. For the Quiet method the equations required to solve the complete problem are in the solution matrices from the outset, but material is removed and added into the model by using material property scaling factors. For the Dead-Alive method, new equations are added into the matrices over the history of the model. Depending on the implementation and the geometry simulated, each method has advantages and disadvantages. These will be described, along with a more technical presentation of these two methods in Chapter 2.

1.3.2 Heat Input

To melt the material during AM, a heat source is required which may be a laser, electronic beam, or an arc torch. It is the heat source that not only makes AM possible, but produces the immense thermal gradients which drive thermal strains, residual stresses, and distortion [51]. The difficulties in modeling the heat source are accurately capturing the size and shape of the input energy volume, and estimating amount of incident heat actually absorbed by the AM component being constructed. A description of several heat input models and their comparative advantages and disadvantages as well a short discussion on how to calibrate the heat source absorption efficiency will be given in Chapter 2.

1.3.3 Thermal Losses

The heat losses which balance the heat input, due to thermal conduction, free and forced convection, and thermal radiation, must be carefully applied to ensure the thermal gradients which drive distortion and stress accumulation are accurately modeled. Modeling the thermal conduction through the part is the central task of a thermal finite element model, which will be detailed in Chapter 2, and tacitly discussed in the remainder of this book. Free convection, stemming from buoyancy effects due to the temperature driven density differences in the atmosphere, can be calibrated or estimated through analytical means. Forced convection, which occurs due to the gas flows built into the various AM machines, is more difficult to calibrate, analytically determine, or measure [40–42]. A basic treatment of convection is presented in Chapter 2, a thorough validation study using various convection treatments is the subject of Chapter 3, and this subject is also touched upon in several other subsequent chapters. Thermal radiation, which can be significant in the region of the melt pool due to the 4th order temperature dependence of this heat transfer mechanism, is also described in Chapter 2. In addition to an explanation of the physical model for thermal radiation, a common FE technique for avoiding excessive non-linearities posed by the 4th order temperature dependence is described.

1.3.4 Distortion and Residual Stress

The application of a high energy heat source to a relatively cool material surface creates large thermal gradients [51,102,103]. These differences in temperature incur thermal expansion. However, the differences in thermal gradients through a single restrained component create exceptional stresses within the part during manufacture. These stresses often exceed the yield strength of the material, which results in the formation of plastic strains; hence AM parts frequently are distorted beyond the specified tolerances for the intended end use of the component. This is why one of the primary motivations for modeling AM processes is to be able to predict, and through iterative simulations, mitigate, undesirable levels of distortion and residual stress. Residual stresses can be alleviated through post-build heat treating. However, once a part distorts excessively there is no recourse to bring the component back into a useable shape. Distortion also has with it the accompanying dangers of cracking, and for LPBF, recoater blade interference, which occurs when the part distorts into the blade which sweeps powder across the building plane prior to the melting of each layer. A discussion of the mechanics of residual stress and distortion formation is included in Chapter 2. Modeling the accumulation of stresses and the distortion history of DED parts are treated in Chapters 8–11. Modeling distortion in LPBF processes is the topic of Chapters 13–15.

1.3.4.1 Coupled Versus Decoupled Models

As just discussed, the thermal behavior during AM processes creates a mechanical response. These two sets of phenomena, the conservation of thermal energy, and the elasto-plastic response, require the solution of differing sets of non-linear equations. To simplify the solution of these problems, AM models separate the two phenomena into thermal and mechanical models. The method in which these two models interact, or their *coupling* can determine both the complexity and accuracy of the complete thermo-mechanical model.

The decoupled or weakly coupled method assumes that the thermal behavior affects the mechanical behavior, but that the mechanical response does not affect the thermal history. Weakly coupled models assume the modeled phenomena are parabolic, where the downstream behavior has no impact upon the upstream behavior. This allows for the thermal history to be simulated first and then the mechanical response to be modeled subsequently.

For a fully or strongly coupled model, this assumption is not made, and instead the mechanical evolution is included in the simulation of the thermal history. Commonly, strongly coupled models will use a sequential scheme, solving a thermal time step, then the associated mechanical time step. If the behaviors are in fact tightly coupled, this could require several iterations of the thermal-mechanical model for each time step until a converged solution is achieved for that time step, before moving on to the next temporal increment.

Technical details of the two coupling methodologies are discussed in Chapter 2. The FE simulations in subsequent chapters all rely upon the assumption of weakly coupled behavior. The success and limitations of this assumption are directly addressed in Chapter 4.

1.3.5 Temperature Dependent Material Properties

If accurate modeling requires the correct application of boundary conditions to determine the heat flux and resulting temperature history through the component, substrate, or other modeled bodies, it follows that the properties that relate to the thermal history and associated mechanical behavior must also be accurate. Material properties can show a large temperature dependence, which must be accounted for to ensure valid models. However the need to model these processes accurately must be balanced out with the desire to simulate these processes in as short time as possible. The resolution of thermal equilibrium requires the determination of the temperature dependent thermal properties which include thermal conductivity, specific heat capacity, density, and emissivity. This is the source of the non-linearity of the thermal model. If the temperature dependence can be eliminated or decreased for any of these properties, it will decrease the non-linearity of the problem and speed up convergence, which ultimately decreases the time required to complete a simulation. However it demands sound engineering judgment, along with experimental and modeling data to determine which material properties are most affected by temperature dependence and which have a negligible effect. This topic will be explored in further detail in Chapter 2.

1.3.6 Microstructural Changes

Like distortion and residual stresses, microstructural changes are inherent in AM processes [61,63,64,66]. After the metal is melted by the heat source it will cool and solidify, and the rate of this solidification largely determines the resulting microstructure. Cooling rates are responsible for not only the resulting grain size, or similar phenomena, but also the precipitation of alloying elements and formation of secondary particles. The cooling rates for AM processes can be several orders of magnitude higher than more traditional casting processes, so that the microstructures which are developed can be vastly different from those heretofore studied and understood. The characterization of AM microstructure is described in Chapter 5 for several commonly used AM metals. Efforts related to the predictive modeling of microstructure are explored in Chapters 6, and 7.

1.3.6.1 *Material Property Changes*

The primary reason to be concerned about the differences in AM microstructure is that the resulting grain size, primary and secondary phases which form can result in materials with similar composition to classically manufactured materials, but with vastly different material properties [104–106]. These changes in material properties have both modeling and practical concerns. As discussed above, the accuracy of thermal-mechanical models rely upon the accuracy of the material properties of the AM component. If these change during deposition, this can be a further source of error or non-linearity, depending on how they are handled by the numeric model. The practical need to put AM built components to actual use means that the changing material properties can be another source of failure. A finished component, even if it is free of high residual stresses and within the desired dimensional tolerances, can be rejected for implementation of its intended end use if its material properties are not within specification, e.g. too low or high of a yield strength. This motivates the desire to know AM microstructure and the ensuing material properties a priori, which is the subject of Chapter 8.

1.3.6.2 *Anisotropy in Material Properties*

As different parts of an AM component will experience differing cooling rates as they are able to conduct, convect, or radiate heat away at different rates, the microstructure within an AM built component is not likely to be homogeneous. This will in turn lead to inhomogeneity and anisotropy in the material properties of AM parts, which have been well documented by experimental studies [17,71]. While post-process heat treatment may limit or remove the anisotropy of the material properties before the component is used, these directional differences in properties may pose a further challenge to the modeler [13]. A brief exploration of these effects is given in Chapter 7.

1.3.6.3 *Phase Transformation*

There are three primary phase transformations an AM model may be forced to solve in order to correctly model the process: liquid–vapor, solid–liquid, and solid state phase transformations. Liquid–vapor phase transformation occurs when the melt pool reaches the boiling point of the deposited metal. There is limited experimental work that indicates this occurs. However, the method of modeling liquid–vapor transformation may be simplified as an evaporative heat loss in the melt pool. The transformation from the solid to liquid and liquid to solid, is an inescapable facet of the AM process. Primarily a model must contend with the extreme non-linearity incurred by the latent heat of fusion or melting. Latent heat is the extra energy required to force a transition between differing states, namely from solid to liquid. When the material cools, this is also the amount of energy released by the molten material upon solidification. The value of latent heat for metals is roughly 100–1000 times the typical values of specific heat capacity. As this occurs at a single temperature, the point where liquid begins to form, latent heat presents an extreme non-linearity in the solution of thermal equilibrium. Improperly applied, latent heat may inhibit convergence in thermal simulations. The importance and methodology of handling the latent heat of fusion to avoid convergence errors is presented in Chapter 2.

While the latent heat of fusion poses problems for thermal model of every material, solid-state phase transformation primarily affects the mechanical behavior and only for select materials. Some metals have two or more common crystallographic structures in their solid state. Different temperatures and pressures can determine which solid phases exist. When the solid state changes phase, the shift to a different crystallographic orientation can produce phase transformation strains, as the microstructure changes both shape and volume. Depending on the material, temperature, and pressure, this can either worsen or mitigate the development of residual stresses and plastic strains in AM components. Modeling solid state phase transformations will be explained in Chapter 2 and applied to actual modeling efforts in Chapters 9–11.

1.3.7 Reducing Simulation Time

Each of the preceding phenomena which may be incorporated into the model may inhibit the timeliness of a computer simulation of AM processes. Modeling material addition may require renumbering the matrices, which can unduly slow down the model. The nonlinearities introduced by boundary conditions, temperature dependent properties, plasticity, and large-scale displacement will require more iterations to converge. However the greatest factor in simulation time is the number of elements required to accurately model the thermo-mechanical phenomena for a specific geometry. Reducing the mesh density by mesh coarsening of a conforming mesh is utilized in Chapters 3, 4, and 8. Nonconforming meshing is used to further reduce simulation time in Chapter 7. A comparison of conforming, nonconforming, and dynamically adaptive meshes for a thermal simulation is explored in depth in Chapter 12. A multi-scale method for the simulation of LPBF processes is applied and validated in Chapters 13–15, which yield thermo-mechanical simulations of part-scale builds in a matter of hours.

1.4 CONCLUSIONS

This chapter has introduced the reader to the vast array of AM processes, the inherent problem of residual stress and distortion, motivated the need for AM process simulation, and described briefly the difficulties to contend with to accurately model these processes. While the numerous modeling challenges outlined above may seem daunting, all of these challenges can be, and many already have been, overcome. This book is a guide for the prospective or practiced AM process modeler to produce or improve useful simulations of AM builds to reduce unwanted behavior, exert greater production control, and eliminate the number of failed parts. This will have the ultimate effect of making AM a useful, economically feasible, and more widely accepted manufacturing technique.

References

[1] Ueda Y, Takahashi E, Fukuda K, Sakamoto K, Nakcho K. Multipass welding stresses in very thick plates and their reduction from stress relief annealing. Trans JWRI 1976;5(2):179–89.

[2] Nickell Robert E, Hibbitt H David. Thermal and mechanical analysis of welded structures. Nucl Eng Des 1975;32(1):110–20.

[3] Rybicki EF, Stonesifer RB. Computation of residual stresses due to multipass welds in piping systems. J Press Vessel Technol 1979;101(2):149–54.

[4] Lindgren Lars-Erik, Runnemalm Henrik, Nässtrom Mats O. Simulation of multipass welding of a thick plate. Int J Numer Methods Eng 1999;44(9):1301–16.

[5] Kar A, Mazumder J. One-dimensional diffusion model for extended solid solution in laser cladding. J Appl Phys 1987;61(7):2645–55.

[6] Griffith ML, Schlienger ME, Harwell LD, Oliver MS, Baldwin MD, Ensz MT, Essien M, Brooks J, Robino CV, Smugeresky JE, Wert MJ, Nelson DV. Understanding thermal behavior in the lens process. Mater Des 1999;20(2):107–13.

[7] Keicher DM, Smugeresky JE. The laser forming of metallic components using particulate materials. JOM 1997;49(5):51–4.

[8] Keicher DM, Smugeresky JE, Romero JA, Griffith ML, Harwell LD. Using the laser engineered net shaping (lens) process to produce complex components from a cad solid model. In: Photonics West'97. International Society for Optics and Photonics; 1997. p. 91–7.

[9] Kruth J-P, Leu Ming-Chuan, Nakagawa Terunaga. Progress in additive manufacturing and rapid prototyping. CIRP Ann-Manuf Technol 1998;47(2):525–40.

[10] Gideon N Levy, Schindel Ralf, Kruth Jean-Pierre. Rapid manufacturing and rapid tooling with layer manufacturing (lm) technologies, state of the art and future perspectives. CIRP Ann-Manufact Technol 2003;52(2):589–609.

[11] Santos Edson Costa, Shiomi Masanari, Osakada Kozo, Laoui Tahar. Rapid manufacturing of metal components by laser forming. Int J Mach Tools Manuf 2006;46(12):1459–68.

[12] Denlinger Erik R, Heigel Jarred C, Michaleris Panagiotis. Residual stress and distortion modeling of electron beam direct manufacturing ti-6al-4v. in: Proceedings of the Institution of Mechanical Engineers, Part B Proc Inst Mech Eng, B J Eng Manuf 2014:0954405414539494.

[13] Shiomi M, Osakada K, Nakamura K, Yamashita T, Abe F. Residual stress within metallic model made by selective laser melting process. CIRP Ann-Manuf Technol 2004;53(1):195–8.

[14] Osakada Kozo, Shiomi Masanori. Flexible manufacturing of metallic products by selective laser melting of powder. Int J Mach Tools Manuf 2006;46(11):1188–93.

[15] Louvis Eleftherios, Fox Peter, Sutcliffe Christopher J. Selective laser melting of aluminium components. J Mater Process Technol 2011;211(2):275–84.

[16] Gao Wei, Zhang Yunbo, Ramanujan Devarajan, Ramani Karthik, Chen Yong, Williams Christopher B, Wang Charlie CL, Shin Yung C, Zhang Song, Zavattieri Pablo D. The status, challenges, and future of additive manufacturing in engineering. Comput Aided Des 2015;69:65–89.

[17] Dinda GP, Dasgupta AK, Mazumder J. Laser aided direct metal deposition of inconel 625 superalloy: microstructural evolution and thermal stability. Mater Sci Eng A 2009;509(1):98–104.

[18] Rosenthal Daniel. Mathematical theory of heat distribution during welding and cutting. Weld J 1941;20(5):220–34.

[19] Ueda Y, Yamakawa T. Analysis of thermal elastic–plastic stress and strain during welding by finite element method. Trans Jpn Welding Soc 1971;2(2):186s–96s.

[20] Goldak J, Chakravarti A, Bibby Malcolm. A new finite element model for welding heat sources. Metall Trans B 1984;15(2):299–305.

[21] Masubuchi K. Analysis of welded structures: Residual stresses, distortion, and their consequences. New York: Pergamon Press; 1980.

[22] Argyris John H, Szimmat Jochen, Willam Kaspar J. Computational aspects of welding stress analysis. Comput Methods Appl Mech Eng 1982;33(1):635–65.

[23] Papazoglou VJ, Masubuchi K. Numerical analysis of thermal stresses during welding including phase transformation effects. J Press Vessel Technol 1982;104(3):198–203.

[24] Kou S, Wang YH. Weld pool convection and its effect. Weld J 1986;65(3):63s–70s.

[25] Chan CL, Mazumder J, Chen MM. Three-dimensional axisymmetric model for convection in laser-melted pools. Mater Sci Technol 1987;3(4):306–11.

[26] Basu Biswajit, Srinivasan J. Numerical study of steady-state laser melting problem. Int J Heat Mass Transf 1988;31(11):2331–8.

[27] Basu Biswajit, Date AW. Numerical study of steady state and transient laser melting problems—I. characteristics of flow field and heat transfer. Int J Heat Mass Transf 1990;33(6):1149–63.

[28] Basu Biswajit, Date AW. Numerical study of steady state and transient laser melting problems—II. effect of the process parameters. Int J Heat Mass Transf 1990;33(6):1165–75.

[29] Mundra K, Debroy T, Zacharia T, David S. Role of thermophysical properties in weld pool modeling. Weld J 1992;71(9):313.

[30] Pardo E, Weckman DC. Prediction of weld pool and reinforcement dimensions of gma welds using a finite-element model. Metall Trans B 1989;20(6):937–47.

[31] Michaleris P, DeBiccari A. Prediction of welding distortion. Weld J Res Suppl 1997;76(4):172s.

[32] Fuhrich T, Berger P, Hügel H. Marangoni effect in laser deep penetration welding of steel. J Laser Appl 2001;13(5):178–86.

[33] Gery D, Long H, Maropoulos P. Effects of welding speed, energy input and heat source distribution on temperature variations in butt joint welding. J Mater Process Technol 2005;167(2):393–401.

[34] Zheng B, Zhou Y, Smugeresky JE, Schoenung JM, Lavernia EJ. Thermal behavior and microstructure evolution during laser-deposition with laser-engineered net shaping: Part II. Experimental investigation and discussion. Metall Mater Trans A 2008;39(9):2237–45.

[35] Anca Andrés, Fachinotti Víctor D, Escobar-Palafox Gustavo, Cardona Alberto. Computational modelling of shaped metal deposition. Int J Numer Methods Biomed Eng 2011;85(1):84–106.

[36] Wang L, Felicelli S. Analysis of thermal phenomena in lens™ deposition. Mater Sci Eng A 2006;435:625–31.

[37] Peyre P, Aubry P, Fabbro R, Neveu R, Longuet Arnaud. Analytical and numerical modelling of the direct metal deposition laser process. J Phys D, Appl Phys 2008;41(2):025403.

[38] Hofmeister W, Wert M, Smugeresky J, Philliber JA, Griffith M, Ensz M. Investigation of solidification in the laser engineered net shaping (LENS™) process. J Minerals Metals Mater Soc 1999;51(7):1–6.

[39] Ye Riqing, Smugeresky John E, Zheng Baolong, Zhou Yizhang, Lavernia Enrique J. Numerical modeling of the thermal behavior during the lens<sup>®<sup> process. Mater Sci Eng A 2006;428(1):47–53.

[40] Heigel Jarred Christopher. Thermo-mechanical model development and experimental validation for directed energy deposition additive manufacturing processes. The Pennsylvania State University; 2015.

[41] Heigel Jarred C, Michaleris Pan, Palmer Todd A. Measurement of forced surface convection in directed energy deposition additive manufacturing. Proc Inst Mech Eng, B J Eng Manuf 2016;230(7):1295–308.

[42] Gouge Michael F, Heigel Jarred C, Michaleris Panagiotis, Palmer Todd A. Modeling forced convection in the thermal simulation of laser cladding processes. Int J Adv Manuf Technol 2015:1–14.

[43] Gouge MF, Michaleris P, Palmer TA. Fixturing effects in the thermal modeling of laser cladding. J Manuf Sci Eng 2017;139(1):011001.

[44] Hoadley AFA, Rappaz M. A thermal model of laser cladding by powder injection. Metall Trans B 1992;23(5):631–42.

[45] Picasso M, Rappaz M. Laser–powder–material interactions in the laser cladding process. J Phys IV 1994;4(C4):C4–27.

[46] Picasso M, Hoadley AFA. Finite element simulation of laser surface treatments including convection in the melt pool. Int J Numer Methods Heat Fluid Flow 1994;4(1):61–83.

[47] de Deus AM, Mazumder J. Two-dimensional thermo-mechanical finite element model for laser cladding. In: Proc. ICALEO, vol. 1996; 1996. p. 174–83.

[48] Hoadley AFA, Rappaz M, Zimmermann M. Heat-flow simulation of laser remelting with experimenting validation. Metall Trans B 1991;22(1):101–9.

[49] Toyserkani Ehsan, Khajepour Amir, Corbin Steve. 3-d finite element modeling of laser cladding by powder injection: effects of laser pulse shaping on the process. Opt Lasers Eng 2004;41(6):849–67.

[50] Costa L, Vilar R, Reti T, Deus AM. Rapid tooling by laser powder deposition: process simulation using finite element analysis. Acta Mater 2005;53(14):3987–99.

[51] Ghosh S, Choi J. Three-dimensional transient finite element analysis for residual stresses in the laser aided direct metal/material deposition process. J Laser Appl 2005;17(3):144.

[52] Ghosh S, Choi J. Modeling and experimental verification of transient/residual stresses and microstructure formation in multi-layer laser aided dmd process. J Heat Trans-T ASME 2006;128(7):662.

[53] Alimardani M, Toyserkani E, Huissoon JP. A 3d dynamic numerical approach for temperature and thermal stress distributions in multilayer laser solid freeform fabrication process. Opt Lasers Eng 2007;45(12):1115–30.

[54] Roberts IA, Wang CJ, Esterlein R, Stanford M, Mynors DJ. A three-dimensional finite element analysis of the temperature field during laser melting of metal powders in additive layer manufacturing. Int J Mach Tools Manuf 2009;49(12):916–23.

[55] Chiumenti Michele, Cervera Miguel, Salmi Alessandro, Agelet de Saracibar Carlos, Dialami Narges, Matsui Kazumi. Finite element modeling of multi-pass welding and shaped metal deposition processes. Comput Methods Appl Mech Eng 2010;199(37):2343–59.

[56] Michaleris P. Modeling metal deposition in heat transfer analyses of additive manufacturing processes. Finite Elem Anal Des 2014;86:51–60.

[57] Kahlen Franz-Josef, Kar Aravinda. Residual stresses in laser-deposited metal parts. J Laser Appl 2001;13:60.

[58] Anca A, Cardona Alberto, Risso José M. 3d-thermo-mechanical simulation of welding processes. Mecánica Comput 2004;23:2301–18.

[59] Ding J, Colegrove P, Mehnen Jorn, Ganguly Supriyo, Sequeira Almeida PM, Wang F, Williams S. Thermomechanical analysis of wire and arc additive layer manufacturing process on large multi-layer parts. Comput Mater Sci 2011;50(12):3315–22.

[60] Denlinger Erik R, Heigel Jarred C, Michaleris Pan, Palmer TA. Effect of inter-layer dwell time on distortion and residual stress in additive manufacturing of titanium and nickel alloys. J Mater Process Technol 2015;215:123–31.

[61] Hassson DF, Zanis C, Aprigliano L, Fraser C. Surfacing of 3.5% nickel steel with inconel 625 by the gas metal arc welding-pulsed arc process. Weld J Weld Res Suppl 1978;(1):1s–8s.

[62] Cortial F, Corrieu JM, Vernot-Loier C. Influence of heat treatments on microstructure, mechanical properties, and corrosion resistance of weld alloy 625. Metall Mater Trans A 1995;26(5):1273–86.

[63] DuPont JN. Solidification of an alloy 625 weld overlay. Metall Mater Trans A 1996;27(11):3612–20.

[64] Xue L, Islam MU. Free-form laser consolidation for producing metallurgically sound and functional components. J Laser Appl 2000;12:160.

[65] Paul CP, Ganesh P, Mishra SK, Bhargava P, Negi J, Nath AK. Investigating laser rapid manufacturing for inconel-625 components. Opt Laser Technol 2007;39(4):800–5.

[66] Rombouts M, Maes G, Mertens M, Hendrix W. Laser metal deposition of inconel 625: microstructure and mechanical properties. J Laser Appl 2012;24(5):052007.

[67] Xu FJ, Lv YH, Xu BS, Liu YX, Shu FY, He P. Effect of deposition strategy on the microstructure and mechanical properties of inconel 625 superalloy fabricated by pulsed plasma arc deposition. Mater Des 2013;45:446–55.

[68] Xing X, Di X, Wang B. The effect of post-weld heat treatment temperature on the microstructure of inconel 625 deposited metal. J Alloys Compd 2014;593:110–6.

[69] Murr LE, Gaytan Sara M, Ramirez DA, Martinez E, Hernandez J, Amato KN, Shindo PW, Medina FR, Wicker RB. Metal fabrication by additive manufacturing using laser and electron beam melting technologies. J Mater Sci Technol 2012;28(1):1–14.

[70] Yadroitsev I, Thivillon L, Bertrand PH, Smurov I. Strategy of manufacturing components with designed internal structure by selective laser melting of metallic powder. Appl Surf Sci 2007;254(4):980–3.

[71] Yadroitsev I, Smurov I. Selective laser melting technology: from the single laser melted track stability to 3d parts of complex shape. Phys Proc 2010;5:551–60.

[72] Ganesh P, Kaul R, Paul CP, Tiwari P, Rai SK, Prasad RC, Kukreja LM. Fatigue and fracture toughness characteristics of laser rapid manufactured inconel 625 structures. Mater Sci Eng A 2010;527(29):7490–7.

[73] Amato K, Hernandez J, Murr LE, Martinez E, Gaytan SM, Shindo PW, Collins S. Comparison of microstructures and properties for a ni-base superalloy (alloy 625) fabricated by electron beam melting. J Mater Sci Res 2012;1(2):p3.

[74] Liu S, Liu W, Harooni M, Ma J, Kovacevic R. Real-time monitoring of laser hot-wire cladding of inconel 625. Opt Laser Technol 2014;62:124–34.

[75] Selcuk C. Laser metal deposition for powder metallurgy parts. Powder Metall 2011;54(2):94–9.

[76] Lin Chun-Ming. Parameter optimization of laser cladding process and resulting microstructure for the repair of tenon on steam turbine blade. Vacuum 2015;115:117–23.

[77] Tikare Veena, Griffith Michelle, Schlienger E, Smugeresky John. Simulation of coarsening during laser engineered net-shaping. Technical report, Albuquerque, NM (United States): Sandia National Labs; 1997.

[78] Ahsan MN, Pinkerton AJ. An analytical–numerical model of laser direct metal deposition track and microstructure formation. Model Simul Mater Sci 2011;19(5):055003.

[79] Casalino G, Campanelli SL, Contuzzi N, Ludovico AD. Experimental investigation and statistical optimisation of the selective laser melting process of a maraging steel. Opt Laser Technol 2015;65:151–8.

[80] Calignano F. Design optimization of supports for overhanging structures in aluminum and titanium alloys by selective laser melting. Mater Des 2014;64:203–13.

[81] Matsumoto M, Shiomi M, Osakada K, Abe F. Finite element analysis of single layer forming on metallic powder bed in rapid prototyping by selective laser processing. Int J Mach Tools Manuf 2002;42(1):61–7.

[82] Patil Rahul B, Yadava Vinod. Finite element analysis of temperature distribution in single metallic powder layer during metal laser sintering. Int J Mach Tools Manuf 2007;47(7):1069–80.

[83] Li Yali, Gu Dongdong. Thermal behavior during selective laser melting of commercially pure titanium powder: numerical simulation and experimental study. Additive Manuf 2014;1:99–109.

[84] Kolossov Serguei, Boillat Eric, Glardon Rémy, Fischer P, Locher M. 3d Fe simulation for temperature evolution in the selective laser sintering process. Int J Mach Tools Manuf 2004;44(2):117–23.

[85] Zhang DQ, Cai QZ, Liu JH, Zhang L, Li RD. Select laser melting of W–Ni–Fe powders: simulation and experimental study. Int J Adv Manuf Technol 2010;51(5):649–58.

[86] Loh Loong-Ee, Chua Chee-Kai, Yeong Wai-Yee, Song Jie, Mapar Mahta, Sing Swee-Leong, Liu Zhong-Hong, Zhang Dan-Qing. Numerical investigation and an effective modelling on the selective laser melting (slm) process with aluminium alloy 6061. Int J Heat Mass Transf 2015;80:288–300.

[87] Foroozmehr Ali, Badrossamay Mohsen, Foroozmehr Ehsan, et al. Finite element simulation of selective laser melting process considering optical penetration depth of laser in powder bed. Mater Des 2016;89:255–63.

[88] Dong Pingsha, Song Shaopin, Zhang Jinmiao. Analysis of residual stress relief mechanisms in post-weld heat treatment. Int J Press Vessels Piping 2014;122:6–14.

[89] Dai Donghua, Gu Dongdong. Thermal behavior and densification mechanism during selective laser melting of copper matrix composites: simulation and experiments. Mater Des 2014;55:482–91.

[90] Baureiß A, Scharowsky T, Körner C. Defect generation and propagation mechanism during additive manufacturing by selective beam melting. J Mater Process Technol 2014;214(11):2522–8.

[91] Khairallah Saad A, Anderson Andy. Mesoscopic simulation model of selective laser melting of stainless steel powder. J Mater Process Technol 2014;214(11):2627–36.

[92] Wang Jin, Yang Mo, Zhang Yuwen. A multiscale nonequilibrium model for melting of metal powder bed subjected to constant heat flux. Int J Heat Mass Transf 2015;80:309–18.

[93] Smith M, Guan Z, Cantwell WJ. Finite element modelling of the compressive response of lattice structures manufactured using the selective laser melting technique. Int J Mech Sci 2013;67:28–41.

[94] Ushijima K, Cantwell WJ, Chen DH. Prediction of the mechanical properties of micro-lattice structures subjected to multi-axial loading. Int J Mech Sci 2013;68:47–55.

[95] Gümrük R, Mines RAW. Compressive behaviour of stainless steel micro-lattice structures. Int J Mech Sci 2013;68:125–39.

[96] Zeng K, Pal D, Gong HJ, Patil N, Stucker B. Comparison of 3dsim thermal modelling of selective laser melting using new dynamic meshing method to ansys. Mater Sci Technol 2015;31(8):945–56.

[97] Dunbar AJ, Denlinger ER, Heigel J, Michaleris P, Guerrier P, Martukanitz R, Simpson T. Experimental in situ distortion and temperature measurements during the laser powder bed fusion additive manufacturing process. Part 1: development of experimental method. Additive Manuf 2016.

[98] Paul R, Anand S, Gerner F. Effect of thermal deformation on part errors in metal powder based additive manufacturing processes. J Manuf Sci Eng 2014;136(3):031009.

[99] Nelaturi Saigopal, Shapiro Vadim. Representation and analysis of additively manufactured parts. Comput Aided Des 2015;67:13–23.

[100] Markl Matthias, Körner Carolin. Multi-scale modeling of powder-bed-based additive manufacturing. Annu Rev Mater Res 2016;46:1–34.

[101] Dunbar Alexander J, Denlinger Erik R, Gouge Michael F, Michaleris Pan. Experimental validation of finite element modeling for laser powder bed fusion deformation. Additive Manuf 2016;12:108–20.

[102] Foroozmehr Ehsan, Kovacevic Radovan. Effect of path planning on the laser powder deposition process: thermal and structural evaluation. Int J Adv Manuf Technol 2010;51(5–8):659–69.

[103] Buchbinder Damien, Meiners Wilhelm, Pirch Norbert, Wissenbach Konrad, Schrage Johannes. Investigation on reducing distortion by preheating during manufacture of aluminum components using selective laser melting. J Laser Appl 2014;26(1):012004.

[104] Zhang Kai, Liu Weijun, Shang Xiaofeng. Research on the processing experiments of laser metal deposition shaping. Opt Laser Technol 2007;39(3):549–57.

[105] Jia Qingbo, Gu Dongdong. Selective laser melting additive manufacturing of inconel 718 superalloy parts: densification, microstructure and properties. J Alloys Compd 2014;585:713–21.

[106] Song Bo, Dong Shujuan, Deng Sihao, Liao Hanlin, Coddet Christian. Microstructure and tensile properties of iron parts fabricated by selective laser melting. Opt Laser Technol 2014;56:451–60.

2

The Finite Element Method for the Thermo-Mechanical Modeling of Additive Manufacturing Processes

Michael Gouge, Pan Michaleris, Erik Denlinger, Jeff Irwin
Product Development Group, Autodesk Inc., State College, PA, USA

INTRODUCTION

This chapter outlines the methodology to numerically solve the equations which describe the thermal and mechanical behavior of AM processes. This discussion follows the method employed by the Netfabb Simulation software which was used to complete the FE validation studies in the rest of the book. However, the modeling approach is general, and may be applied to any AM code.

First, the necessity of using the non-linear finite element method is shown, then a brief explanation of the non-linear FE process is given. The weakly coupled or decoupled modeling method is described, along with reasons for its usefulness and caveats regarding its limitations. A discussion on the thermal and mechanical models is described with the relevant boundary conditions. A treatment of the primary methods to model the addition of material is given, along with the advantages and limitations of each method. A brief discussion on how temperature dependent material properties are applied in the modeling tool is presented. Meshing concerns and methods are then described. Methods of verifying a model is numerically correct by comparing model results to problems with known answers are discussed at length. Finally validation methods are outlined along with error metrics and a discussion of validation criteria.

2.1 A NON-LINEAR FINITE ELEMENT PRIMER

The Galerkin approach is used to convert the governing physics equation to a weak formulation. For a thermal problem, an energy balance is the governing equation, while it is a stress equilibrium for mechanical problems. The weak formulation results in a nodal solution vector $\mathbf{U}$ of either temperatures or displacements, a residual vector $\mathbf{R}$, and a stiffness matrix $d\mathbf{R}/d\mathbf{T}$. From an initial estimate of the solution $\mathbf{U}^0$, the Newton–Raphson method can be iteratively applied:

$$\mathbf{U}^{i+1} = \mathbf{U}^i - \left[\frac{d\mathbf{R}^i}{d\mathbf{U}} \right]^{-1} \mathbf{R}^i \tag{2.1}$$

where the superscripts i and $i + 1$ refer to the previous and current iterations [1]. Equation (2.1) is applied until an appropriate norm of the residual $\mathbf{R}$ is less than a specified tolerance. Each time step takes the solution from the preceding time step as an initial estimate.

Additional details pertaining to the formulation can be found in [2,3].

2.2 THE DECOUPLED MODEL

The decoupled or weakly coupled modeling methodology rests upon the assumption that the relationship between the thermal and mechanical behaviors are unidirectional so that the thermal history affects the mechanical behavior, but that the mechanical behavior has no influence upon the thermal behavior [4]. This assumption is often applicable to AM processes, or at the very least, is a fair approximation. However, this methodology breaks down whenever the distortion of the component changes the system boundaries. An example of this occurrence will be described in Chapter 4.

2.3 MODEL TYPES

A generalized second order, linear partial differential equation [5]

$$A\frac{\partial u^2}{\partial x^2} + B\frac{\partial u^2}{\partial x \partial y} + C\frac{\partial u^2}{\partial y^2} + D\frac{\partial u}{\partial x} + E\frac{\partial u}{\partial y} + Fu + G = 0 \tag{2.2}$$

can be classified as one of three different types of equations:

- Elliptic if $B^2 - 4AC < 0$
- Parabolic if $B^2 - 4AC = 0$
- Hyperbolic if $B^2 - 4AC < 0$

The PDE for conservation of thermal equation is a Parabolic model, which is used to solve the thermal history, while stress equilibrium, which is solved quasi-statically, is elliptic in nature. However, as stress equilibrium uses the thermal history as input, it becomes effectively parabolic, making the system parabolic. Yet this may break down if a change in the mechanical behavior alters the thermal boundary conditions. Examples of when this would occur are:

- Large scale distortion disrupts flow over the part, altering convection boundary conditions
- Distortion changing the geometry or pressure at fixturing surfaces
- Cracking of support structures or component, changing contact surfaces

2.4 THE THERMAL MODEL

The finite element analyses performed in this book are all completed using Netfabb Simulation, a non-linear decoupled 3D transient FE solver. Netfabb Simulation is used to solve a set of governing equations to resolve the thermal history of the direct energy deposition processes. From this thermal history, estimates of the solidification times are produced for the property model. Described below are the governing thermal equations solved by Netfabb Simulation, along with the formulation of the thermal boundary conditions required to resolve the temperature field throughout the history of the deposition process.

2.4.1 Thermal Equilibrium

For a body with constant density, ρ, and an isotropic specific heat capacity, C_p, the governing equation is:

$$\rho C_p \frac{dT}{dt} = -\nabla \cdot \mathbf{q}(\mathbf{r}, t) + Q(\mathbf{r}, t) \tag{2.3}$$

where temperature is T, time is t, $\nabla \cdot$ is divergence, $\mathbf{q}$ is the heat flux, $\mathbf{r}$ is the relative reference coordinate, and body heat source is Q. The necessary initial condition is $T_0 = T_\infty$, where T_∞ is the ambient temperature. A two-part Neumann boundary condition is implemented consisting of the applied heat source and the surface heat losses due to both convection and thermal radiation.

The distribution of heat through the part is described by Fourier's conduction equation:

$$\mathbf{q} = -k \nabla T \tag{2.4}$$

where the isotropic temperature dependent thermal conductivity is k.

To solve these equations it is necessary to have an initial condition, a heat input model, and thermal boundary conditions. The initial condition is set to the temperature of either the ambient or preheating temperature for the substrate or build plate elements. The heat input model may be an applied heat flux or volumetric heat source model, as discussed in the next section. In the following section, the thermal boundary conditions and their numeric implementation are described.

2.4.2 The Heat Input Model

Heat is input into the model either as a surface heat flux boundary condition, or as a volumetric body heat source load.

2.4.2.1 Surface Flux Input Models

Common surface flux inputs include 2D Gaussian distributions, tophats, cones, or tophats with a conical rolloff. All of these distributions may be either circular, or in the general case ellipsoidal.

The most common heat flux input is the 2D Gaussian ellipsoidal distribution [6]:

$$q_{\mathrm{p}} = -k\nabla T = \frac{3P\eta}{\pi ac} \exp\left(-\frac{3x^2}{a^2} - \frac{3\left(z + v_s t\right)^2}{c^2}\right) \quad \text{on surface } S_q \tag{2.5}$$

where P [W] is the heat source power, η is the efficiency, a [m] and c [m] are the width and length respectively of the ellipsoid, v_s [m/s] is the speed of the source, t is time, and S_q is the surface on which the boundary condition is prescribed. The local z direction is parallel to the motion of the heat source, while x is both normal to the motion of the heat source and parallel to the surface S_q.

2.4.2.2 Volumetric Input Models

Common volumetric heat sources include 3D Gaussian distributions, line inputs [7], and various combinations of tophats and cones with different distributions in the depth direction, such as constant up to a certain depth, or linear or exponential decays.

In analogy with the 2D case, the most popular source is Goldak's 3D Gaussian ellipsoidal distribution [6]:

$$Q = \frac{6\sqrt{3}P\eta}{abc\pi\sqrt{\pi}} \exp\left(-\frac{3x^2}{a^2} - \frac{3y^2}{b^2} - \frac{3\left(z + v_s t\right)^2}{c^2}\right) \tag{2.6}$$

where b is the depth of the ellipsoid, x and z are oriented as in Equation (2.5), and y is the depth into the material.

2.4.3 Boundary Losses

During the AM process heat losses may occur due to thermal radiation, free convection, forced convection, or conduction through fixturing bodies. For small builds, e.g. single line experimental depositions, it is common modeling practice to ignore heat losses altogether. This is a fair assumption, as most of the incident heat will be absorbed by the substrate. However the larger the build, the more carefully thermal losses must be modeled to give accurate results. This increased necessity springs from the increased surface area which may radiate or convect heat away. Also longer build times are more likely to heat up the substrate so that fixturing losses may become necessary to model. There is also a modeling consideration which motivates the need for more accurate boundary loss estimations, which is the perpetuation of errors, which scale with volume or time. Temperatures determined at one time step are used as a basis for the next. Any error in a time step then gets passed forward. Again, taking the example of a single bead deposition, the simulation may cover a span of seconds, which mitigates such errors, which would be quickly noticeable in production scale parts.

Differences in machine operation and fixturing can also have outsized effects upon model accuracy. The delivery and shielding gases used in DED processes present a major challenge to accurately model such builds, as described in Chapters 4, 5, and 9. As electron beam systems use vacuum chambers, convection losses are not an issue but this creates a greater pressure to accurately model radiation losses which become the primary mode of surface heat transfer. LPBF machines also employ gas cooling systems whose cooling effect must be approximated in the thermal model to ensure accurate results, as discussed in Chapters 14 and 15. Despite the differences in system, however, the same modeling methodology is used to represent these physical heat loss phenomena.

It is common modeling practice to lump together thermal boundary losses into a single effective heat transfer coefficient:

$$h = h_{\text{free}} + h_{\text{forced}} + h_{\text{rad}} \tag{2.7}$$

where h_{free}, h_{forced} and h_{rad} are the heat transfer coefficients of free convection, forced convection, and linearized radiation, respectively. Then the total heat loss flux is modeled using Newton's Law of cooling:

$$q_{\text{conv}} = h(T_s - T_\infty) \tag{2.8}$$

where q_{conv} is the convective heat flux, h is the heat transfer coefficient, and T_s is the surface temperature.

Convection and radiation losses are described in more detail below.

2.4.3.1 *Convection*

Convection can account for significant amount of the heat losses during AM processes, saving for EBEAM based deposition, which must necessarily take place in a vacuum. The mechanisms of these losses include both free convection and forced convection. Descriptions of these losses, and reasonable rates, are given below.

Free convection, also called natural convection, occurs due to buoyant forces and the density gradient in a quiescent atmosphere. These naturally occur in AM processes when the laser melting heats up the part considerably within a relatively cooler gas environment. This thermal gradient drives free convection. Reasonable rates of free convection are 5–15 W/m^2 °C [8]. These may be easily measured using the lumped capacitance method or modeled using analytical methods.

Forced convection occurs when gas flows over a body with a differing temperature. In DED systems this can be significant, as these typically have a dual gas flow, the propelling gases which deliver the powder, and shielding gases, which cover the melt pool with a non-reactive gas to prevent contamination at high temperatures. Convection values can vary widely, not only from system to system, but depending on the gas flows used. Furthermore, convection from a moving nozzle may have variation itself in 3 dimensions which changes the area of application with time. This can be incredibly difficult to model accurately and depends upon extensive experimental work or CFD. Analytical methods are poor at capturing convection rates in these systems due to turbulence effects and multiple flows. Rough average values may be determined using the lumped capacitance method. For more complete measurements, a 3D map of convection values can be made using a series of hot-wire or hot-film anemometry measurements. Measured peak rates for DED systems range from

40–120 W/m^2 °C [8,9]. However, in absence of measurements a global free and forced convection value of 18 W/m^2 °C is a good first approximation, which has been used for accurate thermal and mechanical models [10].

In LPBF systems it is common to have a small flow of gas streaming across the build chamber, usually far above the build surface. This has a less direct effect upon the component and may be easier to approximate as a single value. Calibration studies show values of 5–20 W/m^2 °C produce accurate models [11].

2.4.3.2 Radiation

The Stefan–Boltzmann law describes the heat flux due to radiation as:

$$q_{rad} = \varepsilon \sigma (T_s^4 - T_\infty^4) \tag{2.9}$$

where ε is surface emissivity and σ is the Stefan–Boltzmann constant, $\sigma = 5.67 \times 10^{-8}$ W/m^2 K^4. Radiation is linearized and treated as an effective heat transfer coefficient through the following means:

$$q_{rad} = h_{rad}(T_s - T_\infty) \tag{2.10}$$

$$h_{rad} = \varepsilon \sigma (T_s + T_\infty)(T_s^2 + T_\infty^2) \tag{2.11}$$

which is incorporated in Equation (2.7). This method reduces convergence errors due to the high temperature dependence of the Stefan–Boltzmann model. These may be further reduced by approximating temperature dependent emissivity as a single average value.

2.4.3.3 Fixturing Losses

Heat loss to fixtures occurs through the mechanism of heat conduction. In DED systems this can apply to clamps or bolts which hold the substrate fast, and the fixturing body, whether a clamping system or a work bench, itself. In LPBF systems fixturing losses occur through the build plate into base of the machine.

Most often models ignore fixturing effects in thermal analyses. This is often a wise and fair modeling choice, as the heat transfer to the fixture is limited. This occurs when the area of the fixture contact is minimal, or the volume of the fixturing device is much smaller than that of the build plate and deposited part. In these cases a modeler may safely ignore fixturing losses, or approximate them using convection losses. However, for systems where the substrate or build plate does heat up, as in the case for thin substrates or larger build volumes, fixturing losses should be addressed either indirectly, via a heightened convection region at the area of contact, or directly. Determining the correct rates of heat transfer is complicated by the roughness of surfaces. Surface roughness limits the direct metal-to-metal contact to peaks, with micro-cavities in between. This leads to a reduction of the rate of heat transfer between two contacting bodies, an effect called contact resistance. The rate of heat transfer is modeled using a reduced effective thermal conductivity, which is known as the gap conductance. Gap conductance values can be approximated in experiments or calibrated using FE simulation and thermal measurements. In cases with significant loses fully coupled models may be required to perform an accurate thermo-mechanical analyses.

When designing experiments for validation, these complications can be and should be avoided. In LPBF systems small builds or large builds with longer inter-layer build times will avoid heating up the substrate. In DED systems using cantilevered systems which limit regions of conduction and depositing further away from the fixturing body or bodies, allows the modeler to safely neglect fixturing effects. However, for real world applications, fixturing and unknown gap conductance complications may be unavoidable. In these cases iterative experimentation may be required to determine proper gap conductance values.

2.4.3.4 Powder Considerations

The properties of the metallic powder are assigned based on powder-solid relationships developed by Sih et al. [12]. The conductivity k_p of the powder consisting of spherical particles can be calculated as follows:

$$k_p = k_f \left[\left(1 - \sqrt{1-\phi}\right) \left(1 + \phi \frac{k_r}{k_f}\right) + \sqrt{1-\phi} \left(\frac{2}{1 - \frac{k_f}{k_s}} \left(\frac{2}{1 - \frac{k_f}{k_s}} \ln \frac{k_s}{k_f} - 1 \right) + \frac{k_r}{k_f} \right) \right] \quad (2.12)$$

where k_f is the thermal conductivity of the argon gas surrounding the particles, ϕ is the porosity of the powder bed, k_s is the conductivity of the solid, and k_r is heat transfer attributed to the radiation amongst the individual powder particles,

$$k_r = \frac{4}{3} \sigma T^3 D_p \quad (2.13)$$

where D_p is the average diameter of the powder particles.

The emission of radiation from the heated porous powder surface is caused by emission from the individual particles as well as from cavities present in the powder bed. The emissivity ε_p of the powder bed can be calculated as:

$$\varepsilon_p = A_H \varepsilon_H + (1 - A_H)\varepsilon_s \quad (2.14)$$

where A_H is the porous area fraction of the powder surface:

$$A_H = \frac{0.908\phi^2}{1.908\phi^2 - 2\phi + 1} \quad (2.15)$$

and ε_H is the emissivity of the powder surface vacancies:

$$\varepsilon_H = \frac{\varepsilon_s \left[2 + 3.082 \left(\frac{1-\phi}{\phi} \right)^2 \right]}{\varepsilon_s \left[1 + 3.082 \left(\frac{1-\phi}{\phi} \right)^2 \right] + 1} \quad (2.16)$$

2.5 THE MECHANICAL MODEL

Following the weakly coupled solution method described above, after the thermal simulation has been completed, which yields a temperature history for each node in the mesh, the quasi-static equilibrium of stress is determined for each time step. The governing stress equilibrium equation is:

$$\nabla \cdot \boldsymbol{\sigma} = \mathbf{0} \tag{2.17}$$

where $\boldsymbol{\sigma}$ is the stress. A constitutive equation is required to relate stress, strain, and the material properties. The mechanical constitutive law is:

$$\boldsymbol{\sigma} = \mathbf{C}\boldsymbol{\epsilon}_e \tag{2.18}$$

where $\mathbf{C}$ is the fourth order material stiffness tensor and $\boldsymbol{\epsilon}_e$ is elastic strain.

2.5.1 Small Deformation Theory

For small deformations, the total strain can be calculated as:

$$\boldsymbol{\epsilon} = \boldsymbol{\epsilon}_e + \boldsymbol{\epsilon}_p + \boldsymbol{\epsilon}_T \tag{2.19}$$

where $\boldsymbol{\epsilon}$, $\boldsymbol{\epsilon}_e$ $\boldsymbol{\epsilon}_p$, and $\boldsymbol{\epsilon}_T$ are the total strain, elastic strain, plastic strain, and thermal strain, respectively.

2.5.1.1 Thermal Strain

The thermal strain is computed as:

$$\boldsymbol{\epsilon}_T = \epsilon_T \boldsymbol{j} \tag{2.20}$$

$$\epsilon_T = \alpha(T - T^{ref}) \tag{2.21}$$

$$\boldsymbol{j} = \begin{bmatrix} 1 & 1 & 1 & 0 & 0 & 0 \end{bmatrix}^T \tag{2.22}$$

where α is the thermal expansion coefficient and T^{ref} is reference temperature.

2.5.1.2 Plastic Strain

The plastic strain is computed by enforcing the von Mises yield criterion and the Prandtl–Reuss flow rule:

$$f = \sigma_m - \sigma_y(\epsilon_q, T) \leq 0 \tag{2.23}$$

$$\dot{\boldsymbol{\epsilon}}_p = \dot{\epsilon}_q \boldsymbol{a} \tag{2.24}$$

$$\boldsymbol{a} = \left(\frac{\partial f}{\partial \boldsymbol{\sigma}}\right)^T \tag{2.25}$$

where f is the yield function, σ_m is Mises' stress, σ_Y is yield stress, ϵ_q is the equivalent plastic strain, and $\boldsymbol{a}$ is the flow vector.

2.5.2 Large Deformation Theory

The governing mechanical stress ($\boldsymbol{\sigma}$) equilibrium equation when the current spacial location of the part $\mathbf{x}$ in a deformed state may vary significantly from the undeformed reference configuration $\mathbf{X}$ state is written as [13]:

$$\nabla_{\mathbf{x}} \cdot \mathbf{P} = 0 \tag{2.26}$$

where the first Piola–Kirchoff stress tensor $\mathbf{P}$ is defined as:

$$\mathbf{P} = J\sigma \cdot \mathbf{F}^{-T} \tag{2.27}$$

where σ is the stress tensor and J is the determinant of the deformation gradient $\mathbf{F}$, which is defined as:

$$\mathbf{F} = \frac{d\mathbf{x}}{d\mathbf{X}} \tag{2.28}$$

The Green strain $\mathbf{E}$ is calculated as:

$$\mathbf{E} = \frac{1}{2}(\mathbf{D} + \mathbf{D}^{T}) + \frac{1}{2}\mathbf{D} \cdot \mathbf{D}^{T} \tag{2.29}$$

where the displacement gradient $\mathbf{D}$ is defined as:

$$\mathbf{D} = \mathbf{F} - \mathbf{I} \tag{2.30}$$

where $\mathbf{I}$ is the identity matrix. For small deformations the total Lagrangian formulation will simplify and equate to the small deformation formulation.

2.5.3 The Quiet Activation Strategy

In the quiet element method, the elements representing metal deposition regions are present from the start of the analysis. However, their properties are assigned so they do not affect the analysis. For heat transfer analyses, the thermal conductivity k is set to a lower value to minimize conduction into the quiet elements, and the specific heat C_p is set to a lower value to adjust energy transfer to the quiet elements:

$$k_{\text{quiet}} = s_k\, k \tag{2.31}$$
$$C_{p\,\text{quiet}} = s_{C_p}\, C_p \tag{2.32}$$

where, k_{quiet} and $C_{p\,\text{quiet}}$ are the thermal conductivity and specific heat used for quiet elements, and s_k and s_{C_p} are the scaling factors used for the thermal conductivity and specific heat, respectively.

The quiet element method has the following advantages:

- It is easy to implement and can be applied to general purpose finite element codes via user subroutines.
- Since the number of elements does not change, the number of equations is constant through the analysis and no additional equation renumbering and solver initialization is needed.

The quiet element method has the following disadvantages:

- If the scaling factors s_k and s_{C_p} are not small enough, the thermal energy conducts into the inactive elements result in errors.
- If the scaling factors s_k and s_{C_p} are too small, the Jacobian may be ill-conditioned.
- Implementation of the quiet element method in modeling additive manufacturing, where most of the analysis domain is composed of quiet elements, may result into long computer runs.

2.5.4 The Inactive Activation Strategy

In the inactive element method, the elements representing metal deposition regions are removed from the analysis and only nodal degrees of freedom corresponding to active elements are considered. Numerical implementation involves computing the element residual and Jacobian for the active elements only and solving for the active nodal degrees of freedom only.

The inactive element method has the following advantages:

- There are no errors or ill-conditioning introduced by scaling factor as in the case of the quiet element method.
- Element residual and Jacobian calculations are performed for active elements only.
- Only the active nodal degrees of freedom are considered at a time resulting into smaller algebraic systems by the Newton–Raphson linearization.

The inactive element method has the following disadvantages:

- The method cannot be easily incorporated into general purpose commercial codes using user subroutines.
- The equation numbering and solver initialization have to be repeated every time elements are activated. This may negate the computational advantage of solving for a reduced active number of degrees of freedom.
- When elements are activated, nodes shared by active elements may not be at the initial temperature which may result into artificial energy being introduced into the model.

2.5.5 The Hybrid Activation Strategy

The hybrid element activation strategy was originally proposed as an improvement over both the quiet and inactive activation strategies [14]. Elements are initially set as inactive. Then, layer by layer elements are switched to quiet, and then switched individually to active, based on the heat input. In this approach, equation numbering and solver initialization is repeated only when each layer is activated resulting into faster computer run times and equivalent results.

2.6 TEMPERATURE DEPENDENT MATERIAL PROPERTIES

Temperature dependent material properties are vital to attain accurate simulations of AM processes. This is due to the large range of temperatures an AM component will experience during production, from room temperature through melting. However these can be difficult to acquire. Some metal manufacturers produce excellent, complete sets of measurements for temperature dependent properties. However novel or highly proprietary alloys are not likely to be as well documented and may required extensive in house testing.

The standard method for implementing temperature dependent properties is to linearly interpolate between measured values. This requires that the experimentalist produced a fine enough resolution of the material property set that can be approximated linearly. At the lowest and highest temperature values, no interpolation can be made. The lower bound is not problematic, as it is unlikely, save for the rare cooled system, for an AM component to be lower than room temperature during manufacture. If there is a gap between the highest temperature and the material's solidus temperature, there may be a temptation to extrapolate. This is inadvisable. While some material properties may have linear, predictable properties across a large temperature range, as does Ti-6Al-4V, more often than not there are large non-linearities at the higher temperatures. It is more prudent to use the highest known temperature at all above temperatures. The exceptions to this are thermal conductivity, yield strength, elongation, and ultimate tensile strength. When attempting to accurately model the melt pool using the FE method, artificially increasing thermal conductivity in the melt temperature region can be used to approximate advective effects in the melt pool. Metals lose most of their strength as the temperature nears melting, so if the data found for an alloy does not show higher temperature strengths it is good practice to add a value 100–200 °C below the solidus, and ascribed to this a very low yield strength, 10% of room temperature strength, and assume perfect plasticity in this region.

2.7 FINITE ELEMENT MESHING FOR ADDITIVE PROCESSES

2.7.1 Mesh Convergence and Heuristics

Using Netfabb Simulation, an 8 node hexahedral mesh is automatically generated from a discrete triangulation of the CAD geometry, i.e. an STL file. For automatic meshes, the elements take an axis-aligned rectangular voxel shape, although the solver is capable of processing any general shaped hexahedron. The proprietary mesh generator is extremely optimized to efficiently deal with the distribution of surface triangles within the 3D space of their enclosed volume, to reduce both memory usage and processing time.

2.7.2 Mesh Convergence and Heuristics

To accurately capture the effects of bending, Netfabb Simulation hex8 meshes require a width of at least 2 elements in the bending direction. However, the geometries of additive manufacturing are typically topology-optimized within lattice type structures, and it may not be obvious to a user what the thinnest cross-section is, or where bending is important.

Therefore, the best practice is to always perform a tripartite mesh convergence study to determine if the mesh is fine enough.

2.7.3 Adaptive Meshing

Netfabb Simulation uses an isotropic h-adaptivity scheme which is used to retain mesh refinement in regions where thermal or displacement gradients are large to maintain accuracy, while coarsening the mesh in other regions to reduce memory usage and processing time. The condensation and recovery method is applied for hanging nodes between varying levels of mesh refinement.

Due to the layer-wise nature of additive manufacturing, a fine mesh is only required within the top few layers. This nature allows for the mesh to be efficiently coarsened in lower regions of the print as the simulation progresses while preserving a close agreement with the original geometry.

2.8 MODEL VERIFICATION

Model verification is the process of ensuring that a numeric model solves the equations it is intended to solve correctly. This is distinctly different from validation, which is often conflated with verification, which attempts to show a model captures a certain physical behavior with experimental measurements. Verification does not rely upon measurements but instead analytical solutions. This is achieved by using the model to solve reduced order, 1D or 2D equations, for which an analytical solution exists. Following this process ensures there are no bugs in the program or mistakes in the underlying method. Patch tests are also useful to ensure proper quadrature. Patch tests may also be combined with an analytical verification method. For more complicated verification the method of manufactured solutions is used to force a solution.

2.8.1 Analytical Solutions

The most straightforward method of verifying an FE solver is by modeling a non-trivial problem with a known analytical solution. Below is a sample of linear and non-linear 1D and 2D problems that can be used to verify an AM code. Verification of the thermal and mechanical behavior is split, as problems combining these two phenomena together go beyond what can be solved analytically.

Verification should work progressively from the simplest to most complex systems. At the outset the task should focus on single element 1D static linear tests. After these are complete verification should move on to non-linear systems and 2D problems at the single element level. These will provide the basis for ensuring functionality of the code. Once this has been established, the same tests should be applied to multiple element systems to verify the discretization process is correct and bug free.

2.8.1.1 1D Static Thermal Analysis

PRESCRIBED HEAT FLUX TO AN ADIABATIC SYSTEM

The governing equation for a steady, linear, adiabatic, 1D heat conduction with an applied flux is:

$$q'' = -k\frac{dT}{dx} \tag{2.33}$$

For a prescribed value of q'' and a set value k, the temperature at the opposite face from the flux, T_L, can be analytically determined by:

$$T_L = \frac{-q''L}{k} + T_0 \tag{2.34}$$

given the arbitrary length L and face 0 temperature, T_0.

PRESCRIBED TEMPERATURE WITH A CONVECTION BOUNDARY CONDITION

The governing equation for steady, linear, 1D heat conduction with a convective flux is:

$$-k\frac{dT}{dx} + h\left(T - T_\infty\right) = 0 \tag{2.35}$$

For any arbitrary known value of h, one can rearrange the above equation to solve for the temperature at the opposing element face, T_L:

$$T_L = \frac{\frac{k}{L}T_0 + hT_\infty}{h + \frac{k}{L}} \tag{2.36}$$

After ascribing arbitrary values to T_0, T_∞, h, k, and L, T_L can be analytically determined. Apply these values in the FE solver and compare to the known solution.

2.8.1.2 Time-Based Thermal Analysis by Lumped Capacitance

The lumped capacitance method can be applied to those situations where a body changes temperature in time, but can be approximated as having no spatial variation in its temperature profile. The generally accepted test for whether this approximation stands is that the Biot number, Bi is low enough, such that:

$$Bi = \frac{hL_c}{k} < 0.1 \tag{2.37}$$

where the conductivity, k, is that of the solid and L_c is the characteristic length, which may be calculated thusly:

$$L_c = \frac{V}{A_s} \tag{2.38}$$

where V is the body's volume and A_s is the external surface area.

As h values typically do not fall below 1, low Biot numbers are generally associated with objects with a large ratio of surface area to volume, like thin plates, or highly conductive materials. The modeler however may choose any combination of these parameters to satisfy the above condition, which allows the lumped capacitance method to be used.

For a body with no radiative losses, exposed to some convective flux, the balance of energy is:

$$-hA_s(T - T_\infty) = \rho V C_p \frac{dT}{dt} \tag{2.39}$$

Letting $\theta = T - T_\infty$ we have:

$$\frac{\rho V C_p}{h A_s} \frac{d\theta}{dt} = -\theta \tag{2.40}$$

Separating and integrating yields:

$$\frac{\rho V C_p}{h A_s} \int_{\theta_i}^{\theta} \frac{d\theta}{\theta} = \int_0^t -dt \tag{2.41}$$

For the initial temperature, difference $\theta_i = T_i - T_\infty$ results in:

$$\frac{\rho V C_p}{h A_s} \ln \frac{\theta_i}{\theta} = t \tag{2.42}$$

This allows to solve for the temperature of the body after a given time under a constant convective flux as follows:

$$T = (T_i - T_\infty) \exp \left[\frac{-h A_s t}{\rho V C_p} \right] + T_\infty \tag{2.43}$$

This equation can be used for the verification of thermal analysis with time, comparing the analytical solution to that determined by the FE code.

2.8.2 The Patch Test

The Patch Test is a classic method of verifying the robustness of a finite-element code. Two nearly identical models are created, one with a regular, identifiable mesh (such as HEX8 in Patran), the other with a skewed mesh, as in Fig. 2.1. In every other way the models are identical e.g. having the same geometry, material properties, and boundary conditions. To verify the code, the two models must return identical results.

2.8.3 The Method of Manufactured Solutions

The method of manufactured solutions is employed to force a solution to a PDE which would otherwise not be solvable by direct analytical methods. Manufactured solutions allow non-linear and 2D sets of equations to be used for verification purposes. For a manufactured solution:

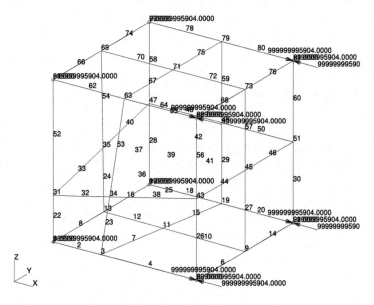

FIGURE 2.1 Patch Test element one dimensional conduction.

1. Write out the governing equation.
2. Assume a polynomial solution of the independent variable to the governing equation. The limiting factor is that the order of the FE element dictates the highest order of polynomial that may be used.
3. Express any dependent variables in terms of polynomials of an order equal or less than that of the independent variable.
4. Substitute the polynomial expressions into the governing equation.
5. Assign arbitrary values to the polynomial coefficients and variables.
6. Solve for the independent variable in terms of the assigned values.
7. Use this value for verification of the FE model.

A simple example may clarify this methodology. Building upon previous analytical solutions, below is the governing equation for transient, 1D conduction with an applied heat flux:

$$\frac{d}{dx}\left(k\left(T\right)\frac{dT}{dx}\right) + Q = \rho C_p \frac{dT}{dt} \tag{2.44}$$

Before attempting higher order verification, 1st order behavior should be tested first, so we will assume a 1st order polynomial solution for a 1st order element, such as hex8:

$$T(x,t) = T_0 + bxt \tag{2.45}$$

The temperature dependent thermal conductivity was also expressed as a 1st order polynomial:

$$k(T) = k_0 + cT \qquad (2.46)$$

When the above expressions are implemented into the above PDE, it becomes:

$$Q = \rho C_p bx - cb^2 t^2 \qquad (2.47)$$

The governing equation for the surface flux is:

$$k(T)\frac{dT}{dx}n + qs = 0 \qquad (2.48)$$

where n is the surface normal at the right face. Using the above functions for $T(x, t)$, the surface heat flux becomes:

$$qs = -kbtn \qquad (2.49)$$

To solve for a temperature, assume values for t, x, k, b, and c. Use these same values in the corresponding FE solution to verify model behavior. This method can be extended to 2D and 3D solutions.

2.9 VALIDATION AND ERROR ANALYSIS

Once a model is verified one can be confident that it solves the equations which form the model correctly. However this is not a guarantor that the model is useful. Validation is the process of showing that a model captures the physics of the actual behavior close enough to be useful. There is some vagueness in the language as the requirements for validity will change from process to process, model to model, or lab to lab. For linear FE codes the generally accepted metric of validity is within 5% of the measurement [15]. For analytical turbulent flow heat transfer models the best solutions are considered accurate if they are ±30% of the measured value [16]. What about the modeling of AM? There is no established standard as of yet, but the literature suggests that ±30% is a fairly acceptable range to deem the results of a simulation accurate. The complexities of the process, with multiple heat transfer modes, melting or even boiling effects, evaporation, powder scatter, changing surface roughness, evolving material properties, phase transformation strains, and imprecise geometric deposition – all present such a challenge that getting within 30% of the measured value is challenging enough for even the most advanced modeling tools. However models may be useful without being quantitatively accurate, so as long as the trends correspond to what is measured during the actual process.

2.9.1 Thermal Error Analysis

For a thermo-mechanical simulation to be accurate, the thermal portion of the model must be accurate as any errors in the temperature field history will propagate into the subsequent mechanical simulation, which will have modeling errors of its own. Due to its primacy in the AM modeling process, validation of the thermal model must take place before validation of the mechanical behavior should be attempted. Below is a short description of common validation techniques and technologies is given.

Thermocouples provide the easiest, most direct, and most accurate temperature measurement methods. However they have their limitations. While it is easy to tack weld thermocouples to metals like steel and nickel alloys, systems like titanium, copper, and aluminum prove more difficult, and may need to be laser welded. Thermocouples are typically attached to the substrate or build plate prior to processing. This practical necessity means that thermocouples are most suited to determining the far field thermal history of an AM process. Yet there are experimental methods that involve threading high temperature thermocouples through holes in a substrate to rest just at the surface, so that measurements at the melt pool may be taken, for rough validation of melt pool models. Note that for this particular method very high sampling rates must be used so as not to lose peak temperatures or cooling rates due to aliasing effects. Additionally, each thermocouple only measures a single location, which is why several thermocouples may need to be used to get an accurate description of the temperature field for validation purposes. A final consideration for those interested in using thermocouples is the possibility of temperature measurement aberrations due to the heat source. Non-physical temperatures spikes have often been observed in thermocouples which are exposed to laser light, within 10–20 mm of the heat source. These measurement aberrations may be mitigated by covering the thermocouple with ceramic beads or epoxy. For electron beam systems temperature aberrations may occur due to the heat source's electro-magnetic effects. Using shielded wires may limit some of these undesirable interactions. However these techniques themselves may alter the temperatures being measured.

Thermal imaging, using either a single pixel pyrometer or an infrared (IR) camera, allow for the recording of temperatures over an evolving surface. Calibration of these systems, however, presents a large challenge to implementing these measurement methods. Pyrometers can be useful over temperatures range widths of 300–1000 °C, depending on the non-linearity of the emissivity of the system [17]. This implies that to fully validate the behavior of an AM process an array of pyrometers is required, to capture temperatures from room temperature through melting. Just like thermocouples pyrometers are also limited to measuring a single location, though the pyrometer may be moved with the laser or build plate to give an Eulerian measurement of temperature. IR cameras are even more difficult to calibrate. For an AM process an IR camera is likely to often capture temperatures from the melting or boiling region all the way down to near room temperature. Camera manufacturers often use their own proprietary calibration methods to aid the experimentalist, but these may not be appropriate for the extreme range posed by AM processes. It is good experimental practice to never use pyrometers or an IR camera without a set of thermocouple measurements to aid the calibration process. This effectively makes thermal imaging a two-step validation process, where first the experimental data itself must be validated, before it can be used to validate the numerical model.

2.9.2 Mechanical Error Analysis

While temperature measurements must necessarily be taken in situ, mechanical analysis can be validated using either post-process or in situ methods. Generally, post-process measurements are adequate to ensure a valid model. However, when a previously validated model shows poor simulation–post-process measurement correlation for a new material or technology, in situ measurements may be required to reveal unexpected behavior.

Thermo-mechanical models are typically used to predict two phenomena: distortion of AM components during fabrication and their residual stresses. Post-process measurements of distortion may be taken using a coordinate measurement machine (CMM) or 3D scans of parts. To determine distortion, the difference between what was built and what was intended to be built must be found. For DED processes, distortion measurements are often limited to the substrate or build plate, due to the imprecise, near-net shape deposition. CMM measurements of the substrate are made before and after processing, then compared to one another to determine the extent of distortion, to compare with the substrate distortion predicted by the model. LPBF builds are far more accurate and precise, and distortion can be easily determined on the components themselves. White light, blue light, or computer tomography (CT) scans of the built part can be compared directly to the warped simulated part. Numeric evaluations of distortion may be calculated by comparing the modeled distortion with the difference between the intended geometry, from the source CAD file, and the scanned part.

Post-process measurements of residual stress are made using one of two general approaches: cutting methods and particle diffraction methods. Cutting methods include blind hole drilling, through hole drilling, and contour methods. These methods involve measuring the strain that is relieved during the material removal process, typically with strain gauges. From the relieved strain the original residual stress may be determined. These methods have low accuracy, roughly ±50 MPa, but are easy and inexpensive to perform [18]. The hole cutting methods are also limited to determining the stress at the surface. A final drawback of the cutting methods are that they are necessarily partially to completely destructive. Diffraction methods, which include X-ray diffraction and neutron diffraction, are non-destructive methods to measure residual stresses. They operate by measuring the diffraction of a high-energy beam through a part, calculating the internal defects from the diffraction, from which the residual stresses may be found. These can be very accurate and yield a field of stresses instead of a single point. However these methods require very specialized equipment, extensive experience to perform properly, and calibration against known stress measurements, typically found via cutting methods.

In situ measurements of distortion can be taken using point measurement devices or strain gauges. Point measurement devices, such as a laser displacement sensor (LDS) or a linear variable differential transducer (LVDT), measure displacement at a single point in a single axis. While this may sound limited, the in situ measurements reveal the cyclic accumulation of distortion which can be useful to validate the cyclic model prediction of distortion, and may expose flaws in the assumed physical behavior, such as neglecting phase transformation effects. However, due to the uniaxial nature of these devices, when there is a significant out of plane distortion the measurements are less useful. This requires some care and perhaps some iterative experiments to find a suitable location to collect model validation data. Moreover, distortion measurements using a LDS or LVDT are practically limited to the substrate or build plate, not the build component itself. This requires using thinner build plates which will more readily distort and constructing housing devices for the experimental equipment, which allows for the measurement of distortion while also protecting the electronics from metal powder and heat. Strain gauges may also be placed on the substrate or build plate prior to deposition. These gauges are better able to capture bending, twisting, or other out of plane distortion modes. However the gauges may need to be placed further away from the heat source as higher temperatures or extreme deformation could cause delamination of strain gauges.

2.10 CONCLUSIONS

This chapter has briefly summarized the methodology of modeling Additive Manufacturing processes using the finite element method. It began with a short treatise on non-linear finite element theory and coupled model schemes. Then the governing equations for the thermal and mechanical systems were discussed, with details of how to apply the modeled heat source, boundary conditions, small or large deformation theory, and microstructural effects. Techniques to simulate material addition were outlined. Non-linear material properties were discussed. Meshing methods and heuristics were described. Methods to verify a model, with detailed examples, were given. Finally a brief description of validation experiments was given, along with a discourse on model validation criteria.

Throughout much of the remainder of this book, these methods will be employed again and again. Details, benefits, and limitations of the proffered modeling techniques will be discussed in detail. The difficulties and benefits of various experimental validation approaches will be shared. It is the hope of these authors that these methods will be of practical use to the reader to construct or improve their own FE modeling attempts.

References

[1] Stoer J, Bulirsch R. Introduction to numerical analysis, vol. 12. Springer Science & Business Media; 2013.
[2] Michaleris P. Modeling metal deposition in heat transfer analyses of additive manufacturing processes. Finite Elem Anal Des 2014;86(0):51–60.
[3] Song J, Shanghvi J, Michaleris P. Sensitivity analysis and optimization of thermo-elasto-plastic processes with applications to welding side heater design. Comput Methods Appl Mech Eng 2004;193(42):4541–66.
[4] Lindgren LE. Finite element modeling and simulation of welding part 1: increased complexity. J Therm Stresses 2001;24(2):141–92.
[5] Haberman R. Elementary applied partial differential equations, vol. 987. Englewood Cliffs, NJ: Prentice Hall; 1983.
[6] Goldak J, Chakravarti A, Bibby M. A new finite element model for welding heat sources. Metall Trans B 1984;15(2):299–305.
[7] Irwin J, Michaleris P. American society of mechanical engineers. A line heat input model for additive manufacturing. J Manuf Sci Eng 2016;138:111004.
[8] Heigel JC, Michaleris P, Palmer TA. Measurement of forced surface convection in directed energy deposition additive manufacturing. Proc Inst Mech Eng, B J Eng Manuf 2016;230(7):1295–308.
[9] Gouge M, Heigel J, Michaleris P, Palmer T. Modeling forced convection in the thermal simulation of laser cladding processes. Int J Adv Manuf Technol 2015:1–14.
[10] Denlinger E, Heigel J, Michaleris P, Palmer T. Effect of inter-layer dwell time on distortion and residual stress in additive manufacturing of titanium and nickel alloys. J Mater Process Technol 2015;215:123–31.

[11] Dunbar AJ, Denlinger ER, Gouge MF, Michaleris P. Experimental validation of finite element modeling for laser powder bed fusion deformation. Additive Manuf 2016;12:108–20.

[12] Sih SS, Barlow JW. The prediction of the emissivity and thermal conductivity of powder beds. Part Sci Technol 2004;22(4):427–40.

[13] Michaleris P, Zhang L, Bhide S, Marugabandhu P. Evaluation of 2D, 3D and applied plastic strain methods for predicting buckling welding distortion and residual stress. Sci Technol Weld Join 2006;11(6):707–16.

[14] Michaleris P. Modeling metal deposition in heat transfer analyses of additive manufacturing processes. Finite Elem Anal Des 2014;86:51–60.

[15] Cook RD, et al. Concepts and applications of finite element analysis. John Wiley & Sons; 2007.

[16] Kreith F, Manglik RM, Bohn MS. Principles of heat transfer. Cengage Learning; 2012.

[17] Inc OE. Non-contact temperature measurement. 2nd edn. Transactions, vol. 1. 1998.

[18] Withers P, Bhadeshia H. Residual stress. Part 1–Measurement techniques. Mater Sci Technol 2001;17(4):355–65.

PART II

THERMOMECHANICAL MODELING OF DIRECT ENERGY DEPOSITION PROCESSES

3

Convection Boundary Losses During Laser Cladding*

Michael Gouge

Product Development Group, Autodesk Inc., State College, PA, United States

3.1 INTRODUCTION

Cladding is the melting of a thin (less than 3 mm) layer of metal upon a surface, to act as a protective coating or to repair damaged surfaces. Laser cladding is a sub-application of Directed Energy Deposition (DED), which can be used to rapidly clad parts [1–4]. The DED process injects metal powder or wire into a melt pool created by a laser beam focused on the part surface. A non-reactive gas flow is used to deliver the metal powder for powder based systems. In both powder and wire systems, an additional shielding gas flow is used to prevent environmental contamination of the melt pool. Each of these gas flows creates localized forced convective cooling on the build surface. In an effort to understand and control the thermal gradients which determine the plastic deformation, residual stresses, and microstructure in DED parts, the convective effects are examined in an ongoing effort to improve DED models [5,6].

The modeling of DED processes was developed from multi-pass welding models [7–14]. Prior DED models which focus on single passes frequently account for natural convection alone [15–18] or neglect convective losses altogether [19,20]. Some thin wall models (less than 3 mm thick) which predict microstructure [21] or control melt pool size [22] also ignore the effect of convection losses but offer no direct validation of thermal history. Han et al. implement natural convection in a two layer model, yet the comparison with experiment shows the model is consistently under-cooled [23].

As the number of passes increases, temperatures at the surface increase, which result in higher rates of heat transfer via conduction, radiation, and convection. As these temperatures increase, the convective heat loss due to the shielding and delivery gases, once negligible, may become significant. These gas flows are typified as impinging jets in the heat transfer

---

* This chapter is based upon the original work: MF Gouge, JC Heigel, P Michaleris, and TA Palmer, "Modeling forced convection in the thermal simulation of laser cladding processes." The International Journal of Advanced Manufacturing Technology 79.1-4 (2015): 307–320. Originally published by ASME.

literature. While the recent work on impinging jets has focused upon Computational Fluid Dynamics (CFD) models [24–27], the foundational literature reports experimental data and algebraic models of convection from various published research sources [28,29]. These empirical functions give the heat transfer coefficient, $h(r)$, as exponential decay functions of radial distance, r, from the nozzle center, which are dependent upon the nozzle-height (H) to nozzle-diameter (D) ratio, H/D, and the nozzle Reynolds number, Re_D. By matching H/D and Re_D the same value for $h(r)$ should be obtained. In DED processes however, the presence of the powder and coaxial flows introduce factors that are not considered in the existing heat transfer studies. Thus, when applying the effect of forced convection to FE models, several different strategies have been explored.

Deus and Mazumder are among the first who use forced convection in a DED model, but they provide few details on its implementation [30]. Vasinonta et al. consider cooling losses from the shielding gases (0.014 W) in their model from which they determine the effect of forced convection to be negligible. However, they offer no experimental validation for the magnitude of these cooling losses, but nevertheless negate convective effects [31]. Dai and Shaw implement a single heat transfer coefficient ($h = 60 \text{ W/m}^2 \text{ K}$) that accounts for cooling by both buoyant effects and the shielding gases [32]. Wen and Shin [33,34] applied unreported values of free and forced convection to their model, while Aggrangsi and Beuth [22] use a free convection value of 5 W/m^2 K and a forced convection value of 10 W/m^2 K. Neither research group validates their models.

Ghosh and Choi conclude that accurate natural and forced convection bounds are necessary to attain good model results [35,36]. The authors do not disclose the values that they use for free or forced convection, but their thermal model is experimentally validated, reporting an error of 13%. Zekovic, Dwivedi, and Kovacevic perform a CFD study to determine the convective bounds for their FE model, yet they do not detail the convection field resulting from their CFD work [37]. However, the inherent caveat when using cascading models is that the assumptions, approximations, and numeric errors from CFD influence the quality of the subsequent thermal FE results.

The effect of implementing experimentally derived convection models in FE simulations of a DED cladding process is investigated. An FE model is developed that is capable of applying convection on the continuously evolving surface of the deposition material. Five possible convection boundary conditions are investigated: no convection, natural convection alone, forced convection measured by lumped capacitance experiments, forced convection extracted from heat transfer literature, forced convection from hot-film anemometry experiments, and a heightened global convection value. These three forced convection models are novel to the DED FE modeling. A comparison is made between using evolving and non-evolving surfaces. Temperature results using the different convection models are compared to experimental measurements. Three methods for developing convection boundary conditions are recommended to improve the accuracy of the FE modeling of any DED process.

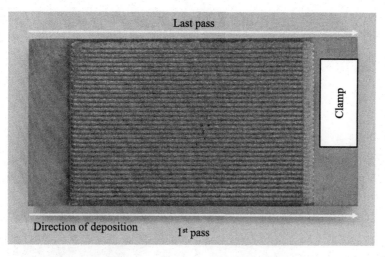

FIGURE 3.1   Presentation of the clad plate. Originally published by ASME in The International Journal of Advanced Manufacturing Technology 79.1-4 (2015): 307–320.

3.2 MODELING APPROACH

The thermal history of the laser cladding process is simulated by solving the conservation of energy equation using the non-linear finite element method with discrete time steps. Temperature dependent properties of Inconel® 625 are implemented. Radiation losses are included using the effective convective coefficient method. Convection losses are approximated using various assumed and measured values of convection described above.

3.3 EXPERIMENTAL PROCEDURES

DED cladding is performed on an Inconel® 625 plate which is 152.4 mm × 76.2 mm × 12.7 mm. A YLR-12000 IPG Photonics fiber laser with 1070–1080 nm wavelength is passed through optical fiber (200 μm diameter) to a colimator (200 mm focal length). After the colimator, the beam is passed through a focusing optic. These experiments used a 200 mm focus optic. This laser is used to melt Inconel® 625 powder (44–149 μm or −100/+325 sieve size). The powder is injected coaxially at a 19.0 g/min through a Precitect® YC50 clad head. The powder is propelled by argon gas flowing at 9 L/min. An additional coaxial 9 L/min stream of argon gas shields the molten pool from oxidation while also protecting the laser optics from powder that might be entrained or reflected from the surface. The clad head maintains a vertical offset of 10 mm above the build, yielding a beam diameter of 4.064 mm on the substrate surface. During deposition, the clad head travels at 10.6 mm/s, while it moves at 31.8 mm/s between deposition passes. Unidirectional cladding is performed in the longitudinal deposition direction, as shown in Figure 3.1. Experimental process parameters are summarized in Table 3.1.

TABLE 3.1 Experimental process parameters for single layer longitudinal cladding

Laser power (kW)	2.4
Laser velocity (mm/s)	10.6
Laser beam diameter (mm)	4.06
Powder feed rate (g/min)	19.0
Pass length (mm)	109
Stepover (mm)	2.03
Number of passes	36

FIGURE 3.2 Cladding experimental setup. Originally published by ASME in The International Journal of Advanced Manufacturing Technology 79.1-4 (2015): 307–320.

The Inconel® 625 substrate is mounted on the fixture shown in Figure 3.2. This fixture was constructed for the measurement of instantaneous deflection by means of a Laser Displacement Sensor (LDS) mounted below the plate. The test fixture has a very small contact area (7.26E-4 m²), minimizing heat loss by means of conduction, essentially isolating surface heat losses to convection and radiation.

Omega® 66-k-30 thermocouples are tack welded to the plate in the four locations shown in Figure 3.3. These thermocouples have an uncertainty of 2.2 °C (or 0.75%, whichever is larger) and a maximum temperature of 1250 °C [38]. TC 1 is located on the center of the bottom surface. TC 2 is located on the bottom surface, at the longitudinal midpoint, near the edge

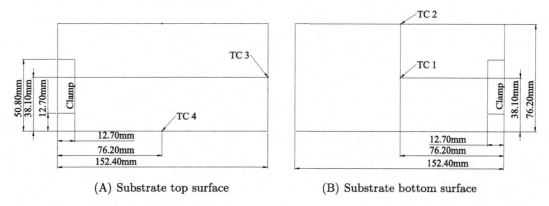

(A) Substrate top surface   (B) Substrate bottom surface

FIGURE 3.3 Thermocouple location schematic. Originally published by ASME in The International Journal of Advanced Manufacturing Technology 79.1-4 (2015): 307–320.

where the first pass is deposited. TC 1 and TC 2 capture the thermal history of the substrate below the cladding region. TC 3 is located on the side nearest the beginning of the deposition passes, at the transverse midpoint, at the top of the plate. This thermocouple, outside of the cladding area, shows the rate at which heat conducts horizontally through the clad plate. TC 4 is located on the longitudinal midpoint on the top of the side nearest the last cladding pass, measuring the temperature as close to the melt pool as possible without incurring thermocouple failure. This array of thermocouples captures the heat transfer through the part along all three cartesian axes. Thus, a simulation which accurately predicts the temperatures at the thermocouple locations can be said to accurately capture the heat conducted through the substrate in the vertical, longitudinal, and transverse directions, the associated temperatures, internal thermal gradients, and near melt pool behavior.

3.4 NUMERIC IMPLEMENTATION

3.4.1 Temporal Discretization

When the laser is on, the time increment is fixed by the laser radius and laser travel speed:

$$t^j = t^{j-1} + \frac{R}{v_w} \tag{3.1}$$

where t^j is the time for the current step, t^{j-1} is the time of the previous step, R is the laser radius, and v_w is the laser travel speed.

3.4.2 Finite Element Mesh

The FE mesh, shown in Figure 3.4, is produced in Patran (MSC Software). The deposition element thickness corresponds to that of the deposited material, while the deposition element

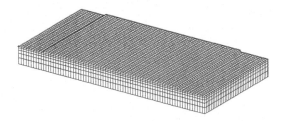

FIGURE 3.4 FE mesh. Originally published by ASME in The International Journal of Advanced Manufacturing Technology 79.1-4 (2015): 307–320.

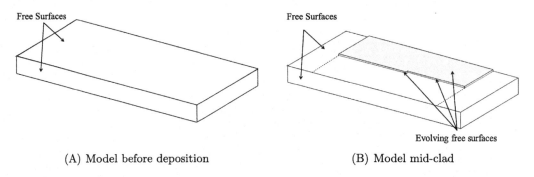

(A) Model before deposition   (B) Model mid-clad

FIGURE 3.5 Evolution of free surfaces during cladding simulation. Originally published by ASME in The International Journal of Advanced Manufacturing Technology 79.1-4 (2015): 307–320.

length and width are equal to the laser radius, yielding a deposited element size of 2.032 mm long, 2.032 mm wide, and 1.168 mm thick. The mesh is coarsened through the thickness of the substrate. The model is comprised of 16302 Hex8 elements and 19929 nodes. A 3 stage mesh convergence study is performed. The meshes are based on 2, 3, and 4 elements per laser diameter. The error between the coarsest and finest mesh is 3.30%, averaged amongst the 4 thermocouples locations, which justifies the use of the coarsest mesh.

3.4.3 The Evolving Free Surface

Surface heat losses from thermal convection and radiation must be applied to the free surfaces of the FE model. While trivial for the thermal modeling of parts with static surfaces, in the modeling of additive processes, which have continuously evolving surfaces, this adds a further complexity. Figure 3.5A depicts the cladding model before deposition, while Figure 3.5B shows the model mid-clad. If the convection bounds were based solely upon the pre-deposition model, then as elements were activated, the convective surfaces would be steadily eliminated, inhibiting heat transfer from the cladding surface elements. If the convective bounds were based solely upon the fully clad model (shown in Figure 3.4) the same error would occur, but in the opposite order, with no convection being applied over the clad area at the beginning of the simulation. Netfabb Simulation implements an algorithm which

TABLE 3.2 Natural convection determination by the lumped capacitance method [39]. Originally published by ASME in The International Journal of Advanced Manufacturing Technology 79.1-4 (2015): 307–320

Property	A36	A6061-T6	Copper
Length (mm)	140	144	140
Width (mm)	139	141	141
Thickness (mm)	6.35	3.12	3.20
ρ (kg/m^3)	7833	2710	8470
C_p (J/kg K)	465	1256	386
k (W/m K)	54	167	386
Bi	1.0E-3	1.7E-4	7.4E-4
h_{avg} (W/m^2 K)	9	9	9

at each time step flags Gauss points which are on the free surface of the mesh then applies the convection and radiation losses stipulated in the model.

Five convective models are investigated. These include no convection, natural convection alone, forced convection from lumped capacitance experiments, forced convection from heat transfer literature, and forced convection measured by hot-film anemometry.

Using no convection is investigated since it is frequently seen in both welding and DED simulation literature [19–22]. Natural convection alone is a common approximation in the DED literature [15–18]. Forced convection from lumped capacitance experiments may offer a low cost method of measuring heat losses under a DED deposition flow, provided they attain adequate accuracy during model validation. Extracting convection data from literature avoids the cost of experimentation entirely, but again must be tested against experimental DED time–temperature data to ensure their accuracy. Hot-film anemometry attains the most detailed measurements of the DED process itself, but may be both laborious and expensive to acquire, and like the previous convection methodologies, the ensuing FE analysis must undergo the same validation process.

3.4.4 No Convection

This method applies neither free nor forced convection to the part. Therefore all heat loss is due to thermal radiation alone.

3.4.5 Natural Convection

To determine the rate of natural convection (also known as free convection), lumped capacitance experiments are completed under quiescent conditions for A36 steel, 6061-T6 aluminum, and commercially pure copper plates [39]. Table 3.2 presents the relevant physical parameters, Biot numbers, and calculated average heat transfer coefficients for the three material types. From the 3 experiments, the mean $h_{avg} = 9$ W/m^2 K.

TABLE 3.3 Forced convection determination by the lumped capacitance method [39]. Originally published by ASME in The International Journal of Advanced Manufacturing Technology 79.1-4 (2015): 307–320

Property	A6061-T6	Copper
Length (mm)	49.4	50.8
Width (mm)	46.8	50.2
Thickness (mm)	3.12	3.20
ρ (kg/m^3)	2710	8470
C_p (J/kg K)	1256	397.5
k (W/m K)	167	386
Bi	1.4E-3	5.3E-4
h_{avg} (W/m^2 K)	66.2	64.8

3.4.6 Forced Convection From Lumped Capacitance Experiments

The lumped capacitance method may also be used to measure an area averaged forced convection. Two preheated test plates are subjected to the cooling effects of both the powder-delivery and shielding gases, from which a lumped capacitance rate of forced convection is determined [39]. The nominal test plate size, 50.8 mm × 50.8 mm, is approximately the average *full width at half maximum* of the smooth and rough anemometry fits. Table 3.3 presents the physical parameters, Biot number, and calculated h_{avg} from these experiments. A mean $h_{\text{avg}} = 65.5$ W/m^2 K over the area of the plate is found by this method.

3.4.7 Forced Convection Based on Published Research

The closest conditions for experimentally determined values of $h(r)$ are found in Saad et al. [40]. As stated above, impinging jet heat transfer regimes may be matched using Re_D, which is the Reynolds number at the nozzle, and H/D, which is the ratio of the height of the nozzle above the surface (H) over the diameter of the nozzle (D). The cladding experiments produce a value of $Re_D = 1594$ (using ambient temperature properties for pure argon gas) and $H/D = 1$. Saad et al. present data for an impinging jet with $Re_D = 1960$ and $H/D = 2$. The authors plot values of $h(r)$ as a function of non-dimensional radial distance $r = x/D$. These values are extracted, re-dimensionalized, and fitted to a curve of the following form:

$$h(r) = Ae^{Br} + h_0 \tag{3.2}$$

where, A is the peak convective value, B is the exponential decay rate, and h_0 is the convective offset. From the work of Saad et al.:

$$h_{\text{Saad}}(r) = 92.8e^{-0.045r} + 0.0 \text{ W/m}^2\text{ K}. \tag{3.3}$$

3.4.8 Hot-Film Anemometry

Hot-film anemometry experiments are completed to determine local h values [39]. Two sets of anemometer measurements are obtained, one for which the sensor is fixed to a smooth

TABLE 3.4 Average roughness measurements. Originally published by ASME in The International Journal of Advanced Manufacturing Technology 79.1-4 (2015): 307–320

Value	Plexiglass	Unclad plate	Anti-skid plate	Clad plate
PV (μm)	4.122	28.99	233.5	164.4
STD PV (μ)	1.443	5.549	78.17	9.655
rms (μm)	0.0071	1.110	33.10	32.85
STD rms (μm)	0.040	0.1141	8.092	1.162
Ra (μm)	0.018	0.967	25.62	26.91
STD Ra (μm)	0.0087	0.775	11.17	1.174

plexiglass plate and one for which the sensor is fixed to a plexiglass plate that is surrounded with anti-skid tape. The smooth plexiglass simulates the convective losses of a virgin substrate surface, while the anti-skid tape experiment represents the surfaced roughed by the deposited material.

The roughness of the clad and unclad surfaces along with their hot-film anemometry experimental representations are quantified through optical profilometry as in reference [41]. Measurement values are tabulated, along with their standard deviations (STD) in Table 3.4. PV is the maximum peak-to-valley distance, rms is the root-mean-squared roughness values, and Ra is the average deviation from the mean. Note that the anti-skid tape was very difficult to measure, with a minimal measurable area. The plexiglass and unclad Inconel® 625 surfaces both have low values of PV (4.122 μm and 28.99 μm respectively), rms (0.0071 μm and 1.110 μm), and Ra (0.018 μm and 0.967 μm). This similarity in surface roughness implies a similarity in boundary layer growth, and thus justifies the use of plexiglass for the convection measurements. The clad and anti-skid tape show agreement as well, particularly in the rms (33.10 μm and 32.85 μm respectively) and Ra (25.62 μm and 26.91 μm) measurements. Again this shows the validity of using the anti-skid tape to represent the clad surface in the hot-film anemometry convection measurements.

The measured values of h are fitted to axisymmetric exponential decay curves of the form in Equation (3.2). The equation for the smooth plate is:

$$h_{smooth}(r) = 75.3 e^{-0.0471r} + 33.7 \text{ W/m}^2 \text{ K} \tag{3.4}$$

while that for the rough surface is:

$$h_{rough}(r) = 69.0 e^{-0.0697r} + 21.7 \text{ W/m}^2 \text{ K}. \tag{3.5}$$

The axisymmetric experimental and curve fit values for both the smooth and rough surfaces are plotted in Figure 3.6. To fit the above curves, the errors are taken, for both the integrated cooling values (Q) and at a point-by-point average, along the observed plane. These errors are:

$$\text{Integrated \% error} = \frac{|Q_{\text{fit}} - Q_{\text{test}}|}{Q_{\text{test}}} \times 100\% \tag{3.6}$$

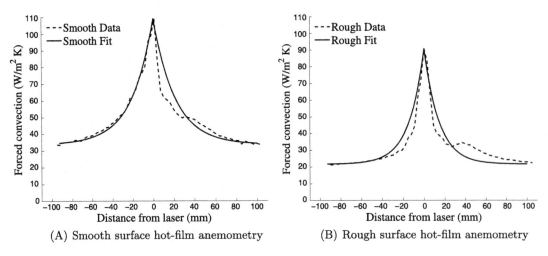

FIGURE 3.6 Hot-film anemometry convection boundary conditions. Originally published by ASME in The International Journal of Advanced Manufacturing Technology 79.1-4 (2015): 307–320.

TABLE 3.5   Anemometer curve fit error. Originally published by ASME in The International Journal of Advanced Manufacturing Technology 79.1-4 (2015): 307–320

Anemometer	Q_{test} (W/m K)	Q_{fit} (W/m K)	Integrated % error	Point-by-point % error
Rough	0.0061661	0.0061664	0.005	10.1
Smooth	0.0096670	0.0096677	0.007	5.17

and

$$\text{Point-by-point \% error} = \frac{1}{n} \sum_{i=1}^{n} \frac{|h_{fit}(r) - h_{measured}(r)|}{h_{measured}(r)} \times 100\% \qquad (3.7)$$

where Q_{fit} is the area under the curve of the exponential fit function and Q_{test} is the area under the curve for the anemometry measurements, in W/m K, n is the number of measured increments for each data set, $h_{fit}(r)$ is the local fit function convection value, and $h_{measured}(r)$ is convection from the two anemometry experiments.

As the exponential decay functions cannot capture the asymmetry of the anemometry data (due to the asymmetry in the gas flows), an integrated convection comparison is used to ensure an equivalent rate of cooling that is applied in the simulations to the measured convection. The point-by-point error gives the average percent of the fit function deviation from the convection measurements. The integrated values and percent errors are presented in Table 3.5. For the three forced convection models, natural convection is applied to those regions unaffected by the gas flows – the bottom and sides of the substrate. The three convection regions for the hot-film method, natural, smooth surface, and rough surface, are depicted in Figure 3.7. On the sides and the bottom, $h_{free} = 9 \text{ W/m}^2 \text{ K}$ is applied. For regions on the surface of the mesh, either in the areas with no deposition or before the activation of deposition elements, the smooth surface convection is applied. Activated deposition element free

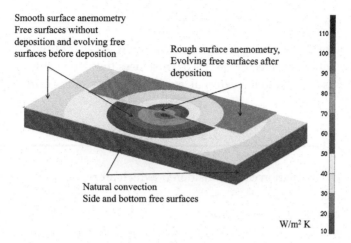

FIGURE 3.7 Hot-film anemometry convection diagram: natural convection alone is applied to the substrate side and bottom surfaces. Smooth surface convection is applied to the top substrate surface. Rough surface convection is applied to the activated deposition elements shown on the back half of the plate. Originally published by ASME in The International Journal of Advanced Manufacturing Technology 79.1-4 (2015): 307–320.

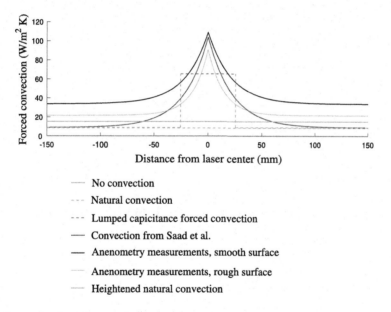

- .... No convection
- - - - Natural convection
- - - - Lumped capicitance forced convection
- .... Convection from Saad et al.
- —— Anenometry measurements, smooth surface
- —— Anenometry measurements, rough surface
- —— Heightened natural convection

FIGURE 3.8 Convection boundary conditions.

surfaces (shown in Figure 3.5) are given the rough surface convection values. This methodology attempts to capture the evolving nature of the clad surface and its impact upon the rate of cooling on the part during cladding. The six convective bounds considered are shown in Figure 3.8.

3.4.9 Heightened Natural Convection

Modeling of thin walls by Denlinger et al. show excellent predictions of distortion using a heightened natural convection [42]. This case uses the same value of 18 W/m² K used in that work which is applied to the external, evolving surface of the cladding and substrate.

3.5 ANALYSIS CASES

There are 7 simulation cases considered, which are:

Case 1: No convection
Case 2: Natural convection alone
Case 3: Forced convection from lumped capacitance experiments
Case 4: Forced convection from heat transfer literature
Case 5: Forced convection measured by hot-film anemometry
Case 6: Forced convection with a non-evolving surface
Case 7: Heightened natural convection

The 7 cases are described below.

3.5.1 Case 1: No Convection

The Case 1 simulation has no convection losses during the entirety of the deposition process.

3.5.2 Case 2: Natural Convection Alone

Case 2 applies $h_{\text{free}} = 9$ W/m² K to all free surfaces at each iteration.

3.5.3 Case 3: Forced Convection From Lumped Capacitance Experiments

For the Case 3 simulation, the average forced convection measured by the lumped capacitance method, $h_{\text{avg}} = 65.5$ W/m² K, is applied to the 50.8 mm × 50.8 mm square region, centered at the laser position, on the top surface of the FE model. All other free surfaces are given the measured natural convection of $h_{\text{free}} = 9$ W/m² K.

3.5.4 Case 4: Forced Convection From Heat Transfer Literature

Case 4 uses a radially decaying function for convection of the type in Equation (3.2), extracted from the heat transfer study by Saad et al. This forced convection is applied to the free substrate top and deposited material surfaces of the part at each iteration. Natural convection, $h_{\text{free}} = 9$ W/m² K, is applied to all other free surfaces.

TABLE 3.6 Analysis cases

Case	Description	A (W/m^2 K)	B	h_0 (W/m^2 K)	Evolving surface
1	No convection	N/A	N/A	0.0	Yes
2	Free convection	N/A	N/A	9.0	Yes
3	Lumped capacitance	N/A	N/A	65.5	Yes
4	Saad et al.	92.8	−0.045	0.0	Yes
5	Anemometry				
	smooth	75.3	−0.0471	33.7	Yes
	rough	69.7	−0.0697	21.7	Yes
6	Anemometry				
	smooth	75.3	−0.0471	33.7	No
	rough	69.7	−0.0697	21.7	No
7	Heightend natural convection	N/A	N/A	18.0	Yes

3.5.5 Case 5: Forced Convection Measured by Hot-Film Anemometry

The hot-film anemometry measurement functions are used in the Case 5 simulation. As stated before, the smooth fit is applied to the undeposited free surfaces on the top of the substrate, the rough fit is applied to the free deposition element surfaces once activated, and natural convection is applied to the sides and bottom of the substrate, as done in Cases 3 and 4 (lumped capacitance forced convection and forced convection from heat transfer literature, respectively).

3.5.6 Case 6: Forced Convection With a Non-Evolving Surface

Case 6 is implemented to show the necessity of applying convection to the evolving free surface at each increment. This case applies convection to a non-evolving surface, as is the case for many commercial FE codes. Case 6 uses the same hot-film anemometry model given in Case 5. The smooth surface data fit is applied to the unclad region of the substrate while the rough surface convection is applied to the external surfaces of the completely clad mesh. Therefore neither free nor forced convective losses are accounted for in the cladding region until the elements at that location have been activated.

3.5.7 Case 7: Heightened Natural Convection

Case 7 applies $h_{\text{free}} = 18$ W/m^2 K to all free surfaces at each iteration.

The 7 cases of the FE analysis presented in this paper are summarized in Table 3.6 according to the form of Equation (3.2).

Area integrated values of cooling are used to compare all of the convection models. Each convection model is integrated over the smallest applied cooling area, the 50.8 mm × 50.8 mm square from Case 3. The method of shells is used to integrate the exponential decay based cooling rates over an equivalent circular area. This method integrates the $h(r)$ values over annular rings. These rings are taken from the midpoints between specified locations. The

TABLE 3.7 Cooling rate comparison

Case	Description	Applied area (m^2)	Integrated cooling rate (W/K)	Evolving surface
1	No convection	N/A	0.000	Yes
2	Natural convection	0.00258	0.023	Yes
3	Forced lumped capacitance	0.00258	0.169	Yes
4	Saad et al.	0.00257	0.106	Yes
5	Anemometry fit			
	smooth	0.00257	0.109	Yes
	rough	0.00257	0.170	Yes
6	Anemometry fit			
	smooth	0.00257	0.109	No
	rough	0.00257	0.170	No
7	Heightend natural convection	0.00258	0.046	Yes

TABLE 3.8 Simulation error for the 7 Analysis Cases

Thermo-couple	Error metric	Case 1: No convection	Case 2: Natural convection	Case 3: Lumped capacitance	Case 4: From Saad et al.	Case 5: Hot-film anemometry	Case 6: Hot-film anemometry	Case 7: Heightened natural convection
Evolving surface		Yes	Yes	Yes	Yes	Yes	No	Yes
TC1	L_2 (°C)	60.3	33.3	13.4	9.07	6.25	21.1	12.7
	% error	16.4	9.00	4.22	5.27	2.80	5.94	3.72
TC2	L_2 (°C)	71.3	41.1	23.1	18.2	13.0	27.0	19.8
	% error	27.5	16.6	9.56	7.50	4.74	11.2	5.03
TC3	L_2 (°C)	50.9	22.3	7.89	7.32	18.6	5.30	
	% error	19.9	10.2	6.49	6.33	12.4	6.24	3.05
TC4	L_2 (°C)	72.9	48.2	29.2	26.3	22.1	41.9	29.5
	% error	19.3	11.2	5.54	4.60	6.03	10.2	5.78

exponential fit functions are integrated over the regions of 1/10th of the anemometry steps ($r \pm 0.3$ mm) for improved accuracy of the numeric integration. The applied areas and integrated rates of cooling of the Cases 1–6 are in given in Table 3.7.

3.6 RESULTS AND DISCUSSION

3.6.1 The Effect of Convection Boundary Conditions

Figure 3.9 illustrates the outcome of the 7 cases of the analysis presented above compared with each of the thermocouple locations. Quantified errors for all 7 Cases at all 4 thermocouples are presented in Table 3.8.

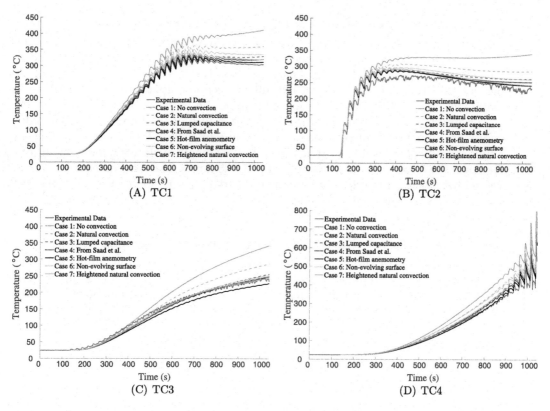

FIGURE 3.9 Comparison of simulated vs experimental temperatures at TC1-4.

3.6.1.1 Case 1: No Convection

Case 1, in which no convection is applied, is the least accurate of all cases considered, with simulated temperatures at all 4 thermocouples exceeding the in situ measurements by 100 °C at least once during each build. The percent errors for Case 1 span 16.4–27.5%, while the L_2 errors are all in excess of 50 °C. The rapid divergence of the temperature histories of Case 1 from the measured temperatures in Figure 3.9 dramatically illustrates the inadequacy of completely ignoring convective losses in DED cladding simulations.

3.6.1.2 Case 2: Natural Convection

Natural convection alone, Case 2, is the least accurate of the models which account for convection. Temperatures are consistently over-predicted, showing the failure of the free convection model to cool the simulated process in a manner consistent with the physical process. As the heat addition continues, errors at the beginning propagate over the remainder of the simulation. Each thermocouple has an L_2 error ranging from 22.3–48.3 °C and percent errors of 9.01–16.6%.

3.6.1.3 *Case 3: Forced Convection From Lumped Capacitance Experiment*

The forced convection from lumped capacitance simulation, Case 3, attains accurate results despite being the simplest description of forced convection. This convection model attains percent errors below 10% for all 4 thermocouples. The L_2 error spans 7.89–29.2 °C while the percent error is 4.22–9.56%. Like the natural convection only method (Case 2), Case 3 consistently overpredicts temperatures, especially at the end of the simulation.

The results from Case 3 underlines the importance of the location at which convection values are applied, not merely the rate of cooling. In Table 3.7 it was shown that Case 3 has an integrated cooling rate nearly identical to the smooth surface anemometry fit of Case 5. However the forced convection is limited to a 50.8 mm × 50.8 mm square, with a convection value lower than both of the hot-film anemometry data sets, so that Case 3 applies a slightly lower cooling rate at a much smaller location than hot-film anemometry based simulation (Case 5).

3.6.1.4 *Case 4: Forced Convection From Heat Transfer Literature*

The plot of Case 4 shows the first radially decaying convection simulation considered, where the value of h diminishes with planar distance from the point of impingement (From Saad et al.). This results in the third most accurate simulation (with percent errors ranging from 4.60–7.50% and L_2 error spans 7.32–26.3 °C), and like the natural convection and forced convection from lumped capacitance results (Cases 2 and 3 respectively), this analysis consistently over-predicts temperatures. This is the reverse of what would be expected from studying both the theoretical and experimental convective research. Impinging jet research suggests that h increases with increasing Re_D and H/D (to some maximum value near $H/D = 7$) [29]. This research informs the a priori assumption that the cooling curve from Saad et al. used here should instead *over-cool* the part during this FE analysis. However, the pure exponential decay, with no convective offset, limits the forced convective effects to a smaller region than for Case 5 (based on the hot-film anemometry measurements) which leads to lower overall cooling. The conclusion may be drawn that the convection behavior of the DED process, comprised of two coaxial, shearing flows and a highly localized heating zone, cannot be perfectly modeled merely by matching dimensionless parameters (Re_D and H/D). However, in lieu of experimental work, approximate boundary conditions may be developed by matching non-dimensional parameters from existent heat transfer literature, which will equal or exceed the accuracy of previous thermal DED modeling work.

3.6.1.5 *Case 5: Forced Convection From Hot-Film Anemometry*

Case 5, based upon the hot-film anemometry experiments, is the overall most accurate simulation, with an L_2 range of 6.25–22.1 °C and a percent error ranging from 2.80–12.4%. Case 5's temperature histories in Figure 3.9 consistently match both the trend and magnitude of the thermocouple measurements. While the lumped capacitance based simulation (Case 3) has similarly small errors, it is unable to match the experimental temperature trends. Again this testifies to the importance of having an FE convection implementation which captures both the total rate of cooling and the location at which the convection is applied during the DED process.

In this vein, there are several improvements that can be made upon the Case 5 simulation. Some of the departure from experiment may be attributable to differences in roughness between the actual part and the representative anti-skid tape surface in the anemometry measurements, which could be accounted for in future anemometry experiments. Comparing Case 5 with Cases 3 and 4 (lumped capacitance forced convection and from Saad's heat transfer study, respectively) at TC 3 indicates that the rate of convection decays somewhat beyond the 100 mm radial distance measured in the anemometry experiments, a distance which again further anemometry experiments or CFD could determine. Other improvements could include modified boundary conditions on the bottom of the plate. The bottom substrate surface will experience re-radiation by the test fixture (Figure 3.2). Furthermore, the substrate bottom, being a downward facing surface with an elevated temperature in a semi-enclosure, will experience a different rate of convection due to buoyant forces and the entrainment of heated gases. Further experimental and modeling work is necessary to pursue these hypotheses.

3.6.1.6 *Case 6: Forced Convection From Hot-Film Anemometry With a Non-Evolving Surface*

Case 6 implements a physically realistic convection model in a non-physically realistic manner. The Case 6 temperature curves presented in Figure 3.9 show that this convection methodology leads to a consistently over-heated cladding simulation, due to the insulating effect of applying convection only upon surfaces that are external in the final mesh. Excluding TC3 (which is significantly overcooled by Case 5) both the percent error and L_2 errors nearly double from Case 5 to Case 6, ranging from 5.94–11.2% and 6.24–41.9 °C respectively. This shows that to accurately capture the thermal behavior of DED cladding in FE simulations, convection must be not only fully representative of the actual cooling rates, but must be applied to an evolving surface just as it occurs during the cladding process.

3.6.1.7 *Case 7: Heightened Natural Convection*

Case 7, the heightened natural convection model, shows remarkable simulation–experiment correlation for the simplicity of this convection modeling method. The temperature curves show a tendency to undercool during deposition, and overcool during the post deposition period. Quantified error analysis indicates it is just behind the Case 4 simulation which used a heat transfer study to produce a 2D convection map. For TC3, at the front of the substrate, Case 7 has the lowest error, just over 3%. This thermocouple is the location that the measured and modeled convection cases consistently overcool, while the no-convection and free convection cases drastically undercool. For studies focused upon bulk behavior, like predictions of deformation and stress, this assumed value of global convection is accurate enough to produce useful model results. However, for researchers concerned with microstructure modeling, cooling rates, or material properties, where capturing the melt pool temperatures accurately is of utmost importance, hot-film anemometry is recommended.

3.7 CONCLUSIONS

Applying experimentally measured convection to a continually evolving mesh surface improves the accuracy of thermal simulations in the FE modeling of DED processes. A methodology for implementing physically representative convection rates in a physically realistic manner for the FE simulation of DED processes has been presented. A total of 7 analysis cases are evaluated: no convection, natural convection alone, forced convection based upon lumped capacitance experiments, forced convection based upon a non-dimensionally similar published heat transfer study, forced convection from hot-film anemometry measurements, hot-film measurement based forced convection using a non-evolving surface, and a heightened natural convection. The accuracy of the simulations is evaluated by comparing simulated temperatures to those of 4 thermocouples placed at 4 locations upon the substrate. A summary of the conclusions is presented below:

1. Ignoring all convective losses has shown to yield inaccurate simulation temperatures.
2. Using natural convection alone is a better approximation than ignoring all convective losses, but is inferior to the models which include forced convection.
3. Forced convection extracted from literature and convection from lumped capacitance experiments can improve the accuracy of thermal DED simulations over the common practice of neglecting forced convection.
4. Using hot-film anemometry measured convection yields the most accurate simulation. This convection model applies an axisymmetric exponentially decaying convection function upon an evolving surface, applying measured smooth surface convection upon unclad free surface elements and measured rough surface convection upon the activated clad elements. This method produces errors of 2.80–12.4%, and L_2 errors between 6.25–22.1 °C.
5. It has been shown that the convection model must be applied to an evolving surface to capture the change of part geometry due to the addition of material during the cladding process.
6. Using a global value of heightened natural convection is shown to be sufficiently accurate for studies not intended to capture melt pool effects.

References

[1] Griffith M, Keicher D, Atwood C, Romero J, Smugeresky J, Harwell L, et al. Free form fabrication of metallic components using laser engineered net shaping (LENS). In: P. solid freeform fab. symps., vol. 9. The University of Texas at Austin; 1996. p. 125–31.
[2] Mazumder J, Choi J, Nagarathnam K, Koch J, Hetzner D. The direct metal deposition of H13 tool steel for 3-D components. JOM-J Min Met Mat S 1997;49(5):55–60.
[3] Griffith M, Schlienger M, Harwell L, Oliver M, Baldwin M, Ensz M, et al. Understanding thermal behavior in the LENS process. Mater Des 1999;20(2):107–13.
[4] Xue L, Islam M. Free-form laser consolidation for producing metallurgically sound and functional components. J Laser Appl 2000;12:160.
[5] Dinda G, Dasgupta A, Mazumder J. Laser aided direct metal deposition of Inconel 625 superalloy: microstructural evolution and thermal stability. Mater Sci Eng A 2009;509(1):98–104.
[6] Rombouts M, Maes G, Mertens M, Hendrix W. Laser metal deposition of Inconel 625: microstructure and mechanical properties. J Laser Appl 2012;24(5):052007.
[7] Ueda Y, Yamakawa T. Analysis of thermal elastic–plastic stress and strain during welding by finite element method. Trans Jpn Weld Soc 1971;2(2):186–96.
[8] Ueda Y, Takahashi E, Fukuda K, Sakamoto K, Nakcho K. Multipass welding stresses in very thick plates and their reduction from stress relief annealing. Trans Jpn Weld Res Inst 1976;5(2):179–89.
[9] Rybicki E, Shadley J. A three-dimensional finite element evaluation of a destructive experimental method for determining through-thickness residual stresses in girth welded pipes. J Eng Mater Technol (United States) 1986;108(2).

[10] Rybicki E, Stonesifer R. Computation of residual stresses due to multipass welds in piping systems. J Press Vessel Technol 1979;101(2):149–54.

[11] Feng Z, Wang X, Spooner S, Goodwin G, Maziasz P, Hubbard C, et al. A finite element model for residual stress in repair welds. TN (United States): Oak Ridge National Lab.; 1996.

[12] Michaleris P, DeBiccari A. Prediction of welding distortion. Weld J Res Suppl 1997;76(4):172s.

[13] Lindgren L, Runnemalm H, Nom M. Simulation of multipass welding of a thick plate. Int J Numer Methods Eng 1999;44(9):1301–16.

[14] Anca A, Fachinotti V, Escobar-Palafox G, Cardona A. Computational modelling of shaped metal deposition. Int J Numer Methods Eng 2011;85(1):84–106.

[15] Hoadley A, Rappaz M, Zimmermann M. Heat-flow simulation of laser remelting with experimenting validation. Metall Trans B 1991;22(1):101–9.

[16] Hoadley A, Rappaz M. A thermal model of laser cladding by powder injection. Metall Trans B 1992;23(5):631–42.

[17] Chin R, Beuth J, Amon C. Thermomechanical modeling of molten metal droplet solidification applied to layered manufacturing. Mech Mater 1996;24(4):257–71.

[18] Han L, Phatak K, Liou F. Modeling of laser cladding with powder injection. Metall Trans B 2004;35(6):1139–50.

[19] Pinkerton A, Li L. An analytical model of energy distribution in laser direct metal deposition. Proc Inst Mech Eng, B J Eng Manuf 2004;218(4):363–74.

[20] Ahsan M, Pinkerton A. An analytical–numerical model of laser direct metal deposition track and microstructure formation. Model Simul Mater Sci 2011;19(5):055003.

[21] Bontha S, Klingbeil N, Kobryn P, Fraser H. Thermal process maps for predicting solidification microstructure in laser fabrication of thin-wall structures. J Mater Process Technol 2006;178(1):135–42.

[22] Aggarangsi P, Beuth J, Griffith M. Melt pool size and stress control for laser-based deposition near a free edge. In: P. solid freeform fab. symps. Austin, TX: University of Texas; 2003. p. 196–207.

[23] Han L, Phatak K, Liou F. Modeling of laser deposition and repair process. J Laser Appl 2005;17(2):89–99.

[24] Cooper D, Jackson D, Launder B, Liao G. Impinging jet studies for turbulence model assessment I. Flow-field experiments. J Heat Trans-T ASME 1993;36(10):2675–84.

[25] Craft T, Graham L, Launder B. Impinging jet studies for turbulence model assessment II. An examination of the performance of four turbulence models. J Heat Trans-T ASME 1993;36(10):2685–97.

[26] Behnia M, Parneix S, Durbin P. Prediction of heat transfer in an axisymmetric turbulent jet impinging on a flat plate. J Heat Trans-T ASME 1998;41(12):1845–55.

[27] Merci B, Dick E. Heat transfer predictions with a cubic model for axisymmetric turbulent jets impinging onto a flat plate. J Heat Trans-T ASME 2003;46(3):469–80.

[28] Gauntner J, Livingood J, Hrycak P. Survey of literature on flow characteristics of a single turbulent jet impinging on a flat plate. Washington, DC. 1970.

[29] Livingood JN, Hrycak P. Impingement heat transfer from turbulent air jets to flat plates: a literature survey; 1973.

[30] de Deus A, Mazumder J. Two-dimensional thermo-mechanical finite element model for laser cladding. In: Proc. ICALEO, vol. 1996; 1996. p. 174–83.

[31] Vasinonta A, Beuth J, Griffith M. Process maps for predicting residual stress and melt pool size in the laser-based fabrication of thin-walled structures. J Manuf Sci E-T ASME 2007.

[32] Dai K, Shaw L. Distortion minimization of laser-processed components through control of laser scanning patterns. Rapid Prototyping J 2002;8(5):270–6.

[33] Wen S, Shin Y. Modeling of transport phenomena during the coaxial laser direct deposition process. J Appl Phys 2010;108(4):044908.

[34] Wen S, Shin Y. Comprehensive predictive modeling and parametric analysis of multitrack direct laser deposition processes. J Laser Appl 2011;23(2):022003.

[35] Ghosh S, Choi J. Three-dimensional transient finite element analysis for residual stresses in the laser aided direct metal/material deposition process. J Laser Appl 2005;17(3).

[36] Ghosh S, Choi J. Modeling and experimental verification of transient/residual stresses and microstructure formation in multi-layer laser aided DMD process. J Heat Trans-T ASME 2006;128(7):662.

[37] Zekovic S, Dwivedi R, Kovacevic R. Thermo-structural finite element analysis of direct laser metal deposited thin-walled structures. In: P. solid freeform fab. symps. Austin, TX: University of Texas; 2005. p. 338–55.

[38] Inc OE. Revised thermocouple reference tables; 2005. Publication Number z204-206.

[39] Heigel J, Michaleris P, Palmer T. Measurement of forced surface convection in directed energy deposition additive manufacturing. Proc Inst Mech Eng, B J Eng Manuf 2016;230(7):1295–308.
[40] Saad N, Douglas W, Mujumdar A. Prediction of heat transfer under an axisymmetric laminar impinging jet. Ind Eng Chem Fundam 1977;16(1):148–54.
[41] Vorburger T. Methods for characterizing surface topography. Tutorials Opt 1992.
[42] Denlinger ER, Michaleris P. Effect of stress relaxation on distortion in additive manufacturing process modeling. Additive Manuf 2016;12:51–9.

CHAPTER

4

Conduction Losses due to Part Fixturing During Laser Cladding*

Michael Gouge

Product Development Group, Autodesk Inc., State College, PA, United States

4.1 INTRODUCTION

Laser cladding is a directed energy deposition (DED) process in which metallic powder or wire feedstock is melted on metallic surfaces using a high power laser to repair damaged surfaces or to enhance surface properties [1–3]. The inherent drawback to laser cladding is that the high energy density heat source can induce high thermal gradients, which, in turn produce thermal strains and residual stresses large enough to drive plastic deformation and distort the part outside its geometric tolerance. The physics of laser cladding then is controlled by the laser heat input and the thermal losses due to radiation, convection, and conduction to the surrounding system. In previous work, both experimental research and Finite Element (FE) modeling has been pursued in order to improve the understanding of the laser cladding process. Heigel et al. used in situ and post process measurements to demonstrate that both magnitude and mode of distortion during laser cladding is dependent upon scan pattern and heat input [4]. Thermal model accuracy was improved by Gouge et al. by implementing measured free and forced convection into the simulation of laser cladding [5]. The current work seeks additional improvements in the accuracy of laser cladding thermal simulations by directly accounting for conduction losses to fixturing bodies in the FE model.

Bodies in contact do not conduct heat as they do through a continuous medium [6]. Contact occurs only where surface peaks extend from one surface to the other, resulting in numerous micro-cavities between points of contact. Heat transfer occurs through the contact points by conduction, and through the micro-cavities by both thermal radiation and conduction through any trapped fluid [7]. The effective rate of conduction through the contacting junction is called the *gap conductance*. Frequently, the inverse of this value is reported as *contact*

---

* This chapter is based upon the original work published by Springer: MF Gouge, P. Michaleris, and TA Palmer. "Fixturing Effects in the Thermal Modeling of Laser Cladding." Journal of Manufacturing Science and Engineering 139.1 (2017): 011001. With the permission of Springer.

resistance [6–8], which has been shown to be a function of contact pressure, contacting materials, temperature, ambient fluid, and the direction of the thermal gradient [6–11]. Theoretical and empirical models of contact resistance have been proposed and validated. Due to the multiplicity of variables, each of these models is process specific which discourages general adaptation into FE models.

Due in part to the complexities just described in the thermal modeling of DED processes, conductive losses to fixtures are most often ignored [12–17]. For small clads on large parts the substrate may have enough thermal capacity to make this a fair model approximation. Once the temperature of the contacting surface rises above ambient temperature this approximation begins to lose its validity, and the thermal interaction between the substrate and the fixture itself becomes important. Several previous modeling efforts have used fixed substrate base temperatures or artificial convection boundaries to mimic conduction losses [18–23]. This body of research is concerned with capturing melt pool behavior, not the thermal history of the larger part, so no validation study was performed showing how the assumed boundary conditions affect the model accuracy. The fixed substrate base temperature technique has also been applied to predictions of distortion in [24,25] which present experimental validation of their models. In both of these works, comparisons of measured and modeled distortion show fair to good accuracy with regards to distortion amplitude, but the model trends fail to fully capture the trends exhibited by the experiments. For such a complex process there are numerous possible reasons for such disagreement between simulation and measurements, but using simplified and uniform thermal boundary conditions is an explicitly stated potential source of error. As of yet, no cladding or direct energy deposition model has attempted to model conductive boundary conditions directly.

However, in the antecedent welding model literature, gap conductance has been well explored. Michaleris and DeBiccari use gap conductance elements between the weld plate and the support structure. Once the gap conductivity is calibrated, excellent agreement between model and experiment is obtained [26]. Ageorges et al. implement gap conductance elements to model contact between interfaces in the modeling of resistance of welded multi-part composite construction. Gap conductance values are calibrated by scaling area-based means of material and air thermal conductivities. Validation is completed by comparing simulated and measured time to melt. The authors found the gap conductance has a minimal effect on the time to melt, most likely due to the very low conductivity of the composite materials [27]. Khandkar et al. use a location dependent effective convection to model conduction losses during friction stir welding. Comparison of simulated and thermocouple measurements show a calibrated contact convection value improves model results [28]. Calibrated effective gap conductance convection has been used to good effect in much of the subsequent friction stir welding modeling literature [29–37]. The effective convection method has also been applied in the modeling of GMAW-based cladding by Mughal et al. [38]. The modeling of laser cladding, which is essentially a series of welding passes, should also benefit from the inclusion of gap conductance elements. Furthermore, the repeated heating cycles of laser cladding processes will result in higher component temperatures, higher thermal gradients at the contact surfaces, and thus higher conduction losses into fixturing. As of yet, no systematic study of incorporating fixturing effects directly has been made within the context of laser cladding.

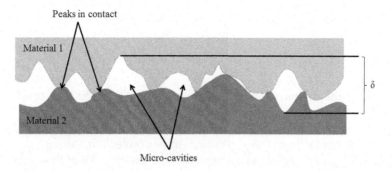

FIGURE 4.1 Illustration of micro-surface contact and micro-cavities.

In this work the effect of accounting for contacting bodies in the thermal modeling of DED cladding processes is investigated. Two experimental designs used in DED research are examined in this study, representing two antipodal extremes of fixturing contact area. One clad is performed on a substrate held in a cantilevered fixture, which has minimal contact area and thus minimal conduction losses. A second substrate is clad while bolted directly to a work bench, which will have substantial conduction losses. The extent of the fixture which should be included in the FE mesh is investigated. An iterative numerical study is undertaken to estimate the gap conductance between the substrate and the fixture. In situ simulated and experimental temperatures are compared when ignoring contact in the model, using contacting bodies with perfect conductance, and using contacting bodies with gap conductance elements. The level of agreement between measured and modeled temperatures demonstrates the impact of gap conductance on the thermal history of laser cladding processes. An analysis of the modeled thermal losses due to convection, radiation, and conduction for each of the two fixturing cases shows the relative importance of conduction losses between a substrate with minimal contact area and cladding a substrate with a significant area of contact.

4.2 MODELING APPROACH

The thermal modeling methodology described in Chapter 2 is used in the following work. This includes the modeling of gap conductance, also known as contact resistance, which is described here in greater detail.

4.2.1 Conduction Losses, Surface Contact, and Gap Conductance

Bodies in contact exhibit imperfect heat transfer by conduction due to the unevenness of real surfaces. The micro-scale roughness of two adjoining surfaces allows true contact to occur only at the peaks extending from the material faces, as shown schematically in Figure 4.1. This condition produces a temperature drop across such junctions as the rate of heat transfer is reduced [6]. From Fourier's Law, the temperature drop between two contacting surfaces is:

$$\Delta T_c = \frac{Q}{k_{gap}\delta} \tag{4.1}$$

where ΔT_c is the intersurface temperature drop, k_{gap} is the gap conductance, and δ is the thickness of the contact region. Gap conductance (k_{gap}) describes the rate at which heat is transferred between adjoining thermally-mismatched surfaces. Heat is transferred through conduction at contacting peaks, conduction or convection through the fluid trapped in the micro-cavities, and thermal radiation within the cavities. All of these behaviors influence gap conductance (k_{gap}). The accepted theoretical approach [7] to gap conductance is to treat these three heat transfer components of gap conductivity, as parallel:

$$\frac{1}{k_{gap}} = \frac{1}{k_{contact}} + \frac{1}{k_{gas}} + \frac{1}{k_{radiation}} \tag{4.2}$$

where $k_{contact}$ is the harmonic mean conductivity of the contacting materials for the peak regions where contact actually occurs, k_{gas} is the conductivity of the gas trapped in the micro-cavities, and $k_{radiation}$ is the thermal radiation within the micro-cavities. Outside very high temperature regimes, the contribution for the effective conduction from gap radiation ($k_{radiation}$) and convection ($k_{convection}$) is negligible, and thus is generally ignored [7].

The peak-to-peak conductivity ($k_{contact}$) of two dissimilar materials [6] in contact is given as:

$$\frac{1}{k_{contact}} = \frac{1}{2}\left(\frac{1}{k_1} + \frac{1}{k_2}\right) \tag{4.3}$$

where k_1 and k_2 are the conductivities of the two materials. However, the total area of actual contact (peak to peak) between two dissimilar materials and the gas filled volume comprised by the micro-cavities is unknown. Without knowledge of the pressure between contacting surfaces, none of the existing models may be implemented. Performing these measurements alters the region and material of contact during cladding, making in situ gap conductance determination impractical. Thus, in the present work the gap conductance will be calibrated based upon experimental results.

4.3 EXPERIMENTAL PROCEDURES

A pair of experiments are performed to illustrate the effect of conductive losses during laser cladding. One substrate is cantilevered in an aluminum fixture, as shown in Figure 4.2. This experiment minimizes the area of thermal contact (2.5% of total surface area, see Table 4.1) and thus limits losses due to conduction at the boundary. An identical substrate is clamped to a work bench using step bolts, as shown in Figure 4.3 using the same process parameters as the cantilevered cladding. With a nominal contact area of 40%, a larger portion of heat should be lost through conduction at the substrate-bench interface.

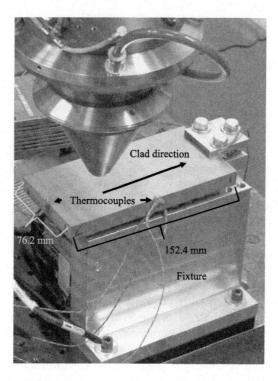

FIGURE 4.2 Cantilevered cladding experimental setup.

TABLE 4.1 Nominal contact area and percentage for the two cladding experiments

Datum	Cantilevered	Bench
Total surface area (mm^2)	29030	29030
Contact area (mm^2)	726	11610
Contact Percent of total surface area	2.50%	40.0%

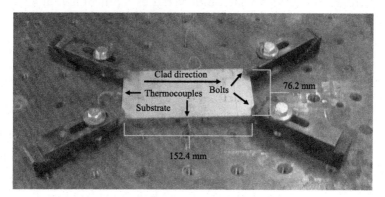

FIGURE 4.3 Work bench cladding experimental setup.

TABLE 4.2 Experimental process parameters for single layer longitudinal cladding

Laser power (kW)	2.4
Laser velocity (mm/s)	10.6
Laser beam diameter (mm)	4.06
Powder feed rate (g/min)	19.0
Pass length (mm)	109
Stepover (mm)	2.03
Number of passes	36

Cladding is performed on Inconel$^{®}$ 625 plates (152.4 mm × 76.2 mm × 12.7 mm) via the DED method. Inconel$^{®}$ 625 is used in for the cladding study as it does not exhibit phase transformation within the cladding time scale [39]. The feedstock is Inconel$^{®}$ 625 powder with a nominal size distribution between 44 and 149 μm. The powder is injected into the melt pool using a Precitec$^{®}$ YC50 clad head at a rate of 19.0 g/min. Argon gas flowing at 9 L/min is used to propel the deposition powder. A second, coaxial argon flow, also 9 L/min, is used to shield the melt pool from contamination. A YLR-12000 IPG Photonics fiber laser (1070–1080 nm wavelength) acts as the heat source. The laser is passed through a 200 μm diameter optical fiber, then a collimator with 200 mm focal length, and finally through a 200 mm focusing optic. The clad head has a 10 mm offset, producing a 4 mm beam diameter at the substrate surface. During deposition, the clad head moves at a speed of 10.6 mm/s. Between passes, the clad head moves at 31.8 mm/s (travel speed). A total of 36 unidirectional passes are deposited on each substrate using the processing parameters of the experiments summarized in Table 4.2. Additional experimental details may be found in [5].

Omega type K thermocouples (TCs) are used to take far field in situ temperature measurements at selected locations on the substrate. These far field temperatures are shown to be sufficient to validate the thermo-mechanical modeling of deposition processes [40]. These thermocouples have an uncertainty of 2.2 °C or 0.75% and a maximum operating temperature of 1250 °C. A NI 9213 module is used to record the data at 20 Hz. Thermocouples are tack welded at 5 locations for the cantilevered experiment and at 4 locations for the bench cladding experiment, as shown the schematic Figure 4.4 and Figure 4.5.

4.4 NUMERICAL IMPLEMENTATION

4.4.1 FE Solver

FE simulations are completed using Netfabb Simulation, a Newton–Raphson based nonlinear FE solver [41]. Deposited elements are activated via the quiet element method which removes and adds elements to the solution matrix by using scaling coefficients, as described in reference [41]. The FE model uses temperature dependent properties. Values between given temperature bounds are determined through linear interpolation. For temperatures beyond the range specified in the property table, the nearest value is used. As for the temporal dis-

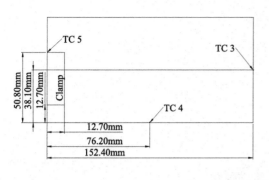

(A) Substrate top surface

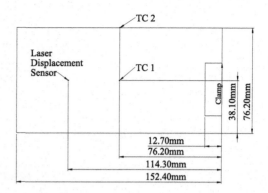

(B) Substrate bottom surface

FIGURE 4.4 Cantilevered thermocouple location schematic.

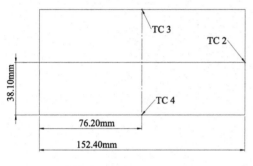

(A) Substrate top surface

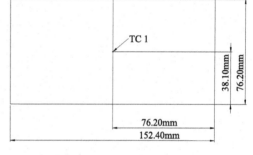

(B) Substrate bottom surface

FIGURE 4.5 Bench thermocouple location schematic.

cretization, each time increment is equal to the time required to move 1 laser radius [40]. The spatial discretization is described below.

4.4.2 The Finite Element Mesh

Three meshes are created using Patran (MSC Software) to simulate the substrate and clad material (Figure 4.6), the substrate in the cantilevering fixture (Figure 4.7), and substrate with the thermally active portion of the work bench (Figure 4.8). Thickness of deposition element is equal to the thickness of the deposition, measured to be 1.2 mm for the present processing conditions [5]. The element length and width are equal to the laser radius (2.0 mm). A tripartite convergence study for the same geometry and process, without either of the contacting bodies, was completed previously by Gouge et al. [5]. Meshes are coarsened through the substrate and fixture to decrease computational time. The clad mesh without the fixture has 19929 Nodes and 16302 Hex 8 elements. The cantilevered mesh is comprised of 30478 Nodes and

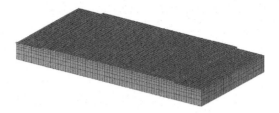

FIGURE 4.6 Substrate only finite element mesh.

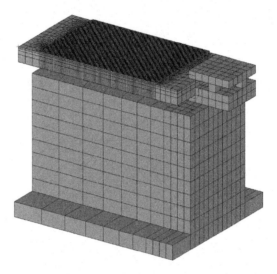

FIGURE 4.7 Cantilevered finite element mesh.

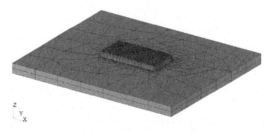

FIGURE 4.8 Work bench finite element mesh.

24754 Hex8 elements, while that for the bench experiment contains 27627 Nodes and 23262 Hex8 elements.

There are features representative of the experimental setup that are not visible in the mesh figures. In Figure 4.7, there are elements representing the stainless steel screws which provide the tension to hold the substrate between the upper and lower portions of the clamp. In Figure 4.8 there are three bolt holes in the work bench (observable in Figure 4.3) underneath the substrate. Holes elsewhere in the workbench are not incorporated into the model, as the bench elements exist solely as a heat sink, and resolving the exact temperature distribution of the bench is not of interest in the present work.

To reduce excessive computational time, it is necessary to minimize the contacting body volume in the mesh. For the cantilevered case, it is trivial to add the entirety of the aluminum fixture. For the work bench based cladding, it is impractical to model the entire volume of the bench. It has been observed during experiments that the work bench surface decays to room temperature within a few substrate lengths from the clad component. The bench area which experiences deviations from room temperature during the build is thus the natural system boundary for attaining thermal equilibrium. An iterative process is used to quantify this boundary. Using the same modeling parameters as in the cantilevered simulation, successive models are run with larger and larger work bench areas. When the exterior nodes are found to be room temperature at all times, the model is said to be sufficient. For the work bench experiment, this condition is met approximately 150 mm from the substrate in each direction.

4.4.3 Convection Model

An axisymmetric convection model based on experimental measurements has been developed, implemented, and validated [5]. This convection model is used for the present simulations as well. A free convection of 9 W/m^2/$^\circ$C was measured by means of the lumped capacitance method. Hot-film anemometry was performed to measure forced convection due to the delivery and shielding gases. From these measurements, two exponential decay models of forced convection, one for the smooth, undeposited surface, and one for the rough, post-deposition surface, have been produced. These models are incorporated into the thermal simulation, calculating the local convection as a function of radial distance from the laser location at each time step. The complete details of the application, values, and validation of the convection model are described in previous work [5]. A plot of the applied forced convection as an axisymmetric function of distance from the simulated heat source is presented in Figure 4.9.

4.4.4 The Gap Conductance Model

Gap conductance is modeled as a thin layer of elements at the substrate-fixture contact. This approach relies upon the following approximations to be used in the current decoupled model:

1. Gap conductance is uniform and identical for each contacting face.
2. Gap conductance is isotropic.
3. Gap conductance is constant over the deposition history. Thus pressure, temperature, and deformation effects are ignored.
4. Conductivity is the only gap conductance element property altered. Specific heat and density are the same as the fixturing material at the given temperature.
5. Bolt contact is not subject to gap conductance.

For the cantilevered model (illustrated in Figure 4.10), there are two horizontal contact faces, on the top and bottom of the substrate, and a single vertical contact face, where the substrate

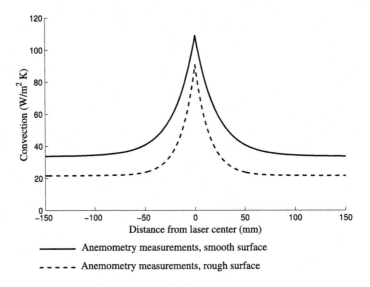

FIGURE 4.9 Forced convection as an axisymmetric function from the laser heat source center.

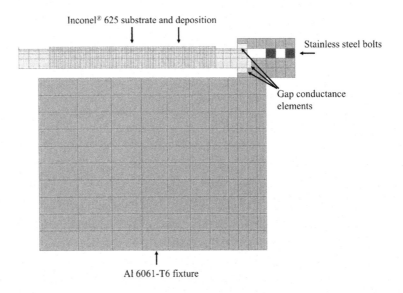

FIGURE 4.10 Cantilevered substrate gap conductance finite element mesh.

butts against the cut-away edge of the lower fixture face. Three thicknesses of the contact resistance elements are modeled to ensure mesh convergence, equal to one-half, one-quarter, and one-eighth of the original element thickness. Differences in the model temperatures between the three thicknesses are negligible, thus the coarsest mesh is used, using the half-size elements as shown in Figure 4.10.

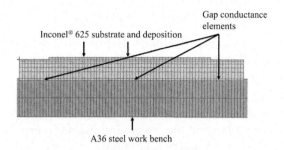

FIGURE 4.11 Workbench bolted substrate gap conductance finite element mesh.

TABLE 4.3 Gap conductance values

Inconel® 625-Aluminum 6061-T6 contact						
Temperature (°C)	$k_{contact}$ (W/m °C)	5% $k_{contact}$ (W/m °C)	1% $k_{contact}$ (W/m °C)	R_c (°C m²/W)	5% R_c (°C m²/W)	1% R_c (°C m²/W)
21	18.5	0.925	0.185	17.0	340	1700
98	20.7	1.93	0.207	15.2	304	1520
201	23.5	1.17	0.235	13.4	267	1340
316	26.4	1.32	0.264	11.9	238	1190
428	29.3	1.47	0.293	10.7	214	1070
571	33.5	1.68	0.335	9.38	188	938
Inconel® 625-A36 Steel contact						
Temperature (°C)	$k_{contact}$ (W/m °C)	5% $k_{contact}$ (W/m °C)	1% $k_{contact}$ (W/m °C)	R_c (°C m²/W)	5% R_c (°C m²/W)	1% R_c (°C m²/W)
25	16.6	0.829	0.166	47.5	949	4750
800	23.1	1.16	0.231	34.0	681	3400

For the bench model (shown in Figure 4.11), the sole contacting face considered is that between the substrate and the work bench. Contact at the step bolt faces is not considered, as this area is negligible compared to the area of contact at the substrate base. In the fixture mesh, the elements in the contact region are quite coarse, so the gap conductance elements are of the same order of magnitude as the deposition elements, resulting in 1.27 mm thick elements.

The approximate value of gap conductance for each experimental setup is calibrated through successive thermal simulations. First, the initial conduction values ($k_{contact}$) used are the maximum expected from perfect contact with no micro-cavities, according to equation (4.3). Subsequent simulations estimate gap conductance as a percentage of this maximum value. Calibration indicates gap conductance for the cantilevered model asymptotes at 1% k_{gap} and for the bench model, lies between 1% k_{gap} and 5% k_{gap}. The base and calibrated gap conductivities along with their equivalent contact resistances (R_c) are listed in Table 4.3. The calibrated contact resistivities are in the range for general metal–metal contact (1×10^2–5×10^4 (°C m²/W)) given by Gmelin et al. [11].

4.4.5 Modeling Assumptions and Approximations

Implicit in the formulation of the thermal equilibrium used in this model, application of the boundary conditions and their numeric implementation are several assumptions and modeling approximations which are explicitly listed below:

1. Density remains constant with temperature for all materials.
2. Material properties between reported temperatures change linearly.
3. Material properties beyond the lowest and highest reported temperatures are constant.
4. The laser absorption coefficient is constant.
5. The free convection coefficient is constant with surface orientation, material composition, and temperature.
6. Thermal emissivities are constant with surface orientation and temperature.
7. Thermal radiation between modeled surfaces are assumed to be negligible.
8. Reradiation from both surroundings are assumed to be negligible.
9. The ambient temperature remains constant during the deposition and cool down process.
10. Thermal and mechanical behaviors are decoupled, such that mechanical behavior due to thermal strains does not alter the thermal history during deposition.

4.5 RESULTS AND DISCUSSION

A comparison between the measured and modeled temperatures are presented for both experimental setups. Each set of experimental temperatures is compared with four approximations of conduction behavior:

- No fixture uses the substrate only mesh and thus has no conduction losses.
- $k_{contact}$ is given the theoretical maximum conductivity described by Equation (4.3).
- 5% $k_{contact}$ is 5% of the theoretical maximum value of conductivity.
- 1% $k_{contact}$ is 1% of the theoretical maximum value of conductivity.

These two values of $k_{contact}$ are used as they approach the measured temperatures for both fixturing cases.

4.5.1 Cantilevered Fixture

Figure 4.12 shows a comparison between the measured temperatures and the four simulated contact conditions at selected locations on the substrate for the cantilevered fixture. For TC1-TC4, differences in the plotted temperature histories and quantified errors are negligible between the four contact approximation methods. Prior to incorporating the fixture the simulation slightly overpredicts measured temperatures, particularly during the post-deposition cool down period. Adding the perfect contact fixture in the $k_{contact}$ case, the model unpredicts the experimental trends as an excess of heat is transferred into the aluminum fixture. Calibration of $k_{contact}$ mitigates the underprediction of thermocouple temperatures. However, the changes from the No Fixture to the 1% $k_{contact}$ are minimal, exhibiting no appreciable change

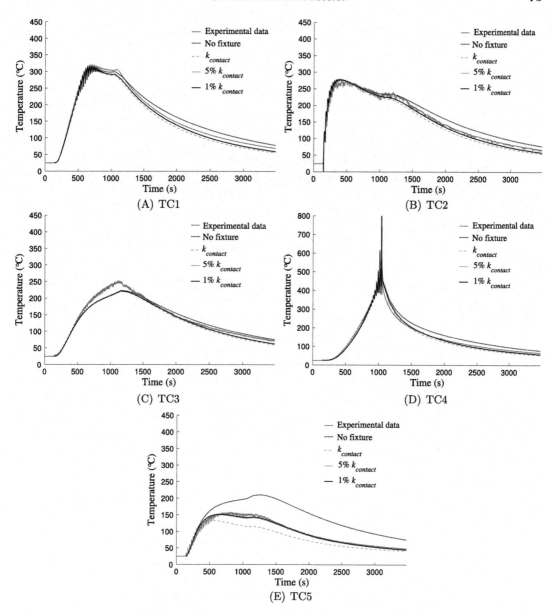

FIGURE 4.12 Comparison of simulated vs experimental temperatures at TC1-5, cantilevered cladding.

in temperature trends. This indicates that conduction losses and gap conductance effects are negligible further away from the region of contact. At TC5, near the region of fixture contact, significant improvement in the model trend can be observed using the fixturing approximations. Ignoring the fixture, as is common practice, leads to model temperatures which

TABLE 4.4 Cantilevered simulation error

TC	No fixture	$k_{contact}$	5% $k_{contact}$	1% $k_{contact}$
TC1	2.22%	3.21%	2.84%	2.80%
TC2	4.49%	4.93%	4.59%	4.56%
TC3	6.48%	6.63%	6.56%	6.56%
TC4	6.04%	6.79%	6.42%	6.38%
TC5	20.5%	13.4%	6.90%	6.49%

TABLE 4.5 Bench simulation error

TC	No fixture	$k_{contact}$	5% $k_{contact}$	1% $k_{contact}$
TC1	91.2%	29.4%	11.2%	44.6%
TC2	77.7%	35.8%	26.1%	9.70%
TC3	59.5%	22.5%	15.1%	11.4%
TC4	102%	38.0%	21.4%	13.6%

approach 100° over the measured values. Implementing the perfect contact fixture results in underpredicting temperatures near the clamp, yet these are far more accurate in both value and trend than neglecting contacting losses altogether. Calibration of $k_{contact}$ show that both the 1% and 5% cases capture the measured temperature history almost perfectly. This thermocouple clearly illustrates the errors that may occur by assuming losses due to conduction are negligible and that the gap conductance calibration methodology presented in this work may significantly improve experiment–model accuracy.

Table 4.4 reports the percent error for the four gap conductance conditions applied to the cantilevered fixture simulation. Quantified error decreases from 20.5% to 13.4% at TC5 between the no fixture and $k_{contact}$ simulation. Calibration of the gap conductance shows further improvement at TC5, with the lowest error being captured by the 1% $k_{contact}$ simulation. From the original no fixture model to the 1% $k_{contact}$ simulation, the error decreases from 20.5% to 6.49%, which is the same level of accuracy attained by the other thermocouples. This increase in model accuracy will improve the accuracy of subsequent simulations of mechanical or microstructural behavior.

4.5.2 Bench Clamped Fixture

Figure 4.12 shows a comparison between the measured temperatures and the four simulated contact conditions at selected locations on the substrate for the bench clamped fixture. Without accounting for the fixture, the bench cladding simulation results are significantly higher than the measured temperatures at each location by up to 200°. Once the bench is incorporated, as shown in the $k_{contact}$ time–temperature plot, the ability of the model to accurately capture the measured temperature trend is enhanced. However, the model significantly underpredicts temperature due to excessive conduction losses resulting from assuming perfect contact. The 5% and 1% $k_{contact}$ simulations improve upon this accuracy further while neither one perfectly reflecting the measurements, as further evidenced by looking at the quantified error in Table 4.5.

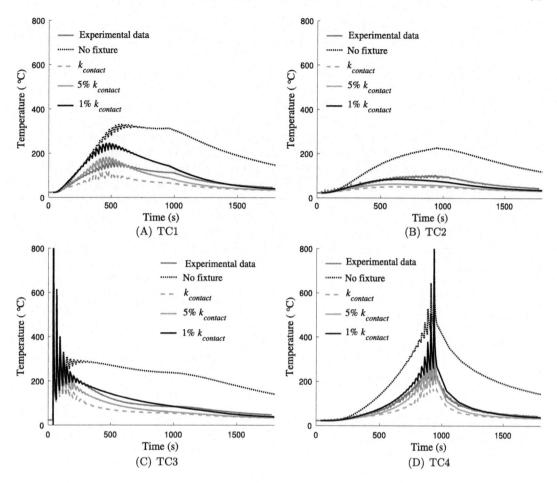

FIGURE 4.13 Comparison of simulated vs experimental temperatures at TC1-4, bench cladding.

As shown in Table 4.5 the implementation of gap conductance significantly improves model–measurement correlation. The quantified error decreases from a range of 60–102% for the simulation without the fixture to a range of 10–45% between the 1% and 5% $k_{contact}$ models. Unlike the cantilevered simulations, neither the 1% or 5% $k_{contact}$ produce superior results at all temperature measurement locations for the duration of the experiment. It is known that plastic deformation occurs during the deposition process. Thus it is suggested that the deformation of the substrate alters the contact pressure and therefore the gap conductance during deposition. A closer examination of the simulated thermal histories provides additional insights.

Figure 4.13A, which depicts the temperature history at the center of the substrate base (TC1), exhibits the strongest evidence for the importance of accounting for contact pressure in gap conductance during the simulation of laser cladding. The 5% $k_{contact}$ simulation matches the measurement trend fairly well until around the 600 s mark, at which point the simulation temperatures fall below those of the experiment. Similarly in Figure 4.13B, measured on the top edge of the substrate at the transverse midpoint (TC2), the 1% $k_{contact}$ model temperatures are almost identical to that of the measurement up until the 600 s, after which the model underpredicts experimental temperatures. These two temperature histories indicate that the contact pressure is changing with both time and location. For example, there is a higher contact conductance at the center of the plate (TC1) than at the top edge midpoint (TC2), during the first two-thirds of the deposition. This indicates that prior to deposition the pressure is higher at the center than the edges. Additionally, the underprediction of temperatures past 600 s shows that pressure decreases for both locations during deposition.

The temperature histories shown in Figure 4.13C and Figure 4.13D, on the base of the plate at the longitudinal midpoint (TC3) and opposite of TC3 but on the top of the substrate (TC4), respectively, reveals a contravening trend. As in TC2, also along the perimeter of the substrate, the 1% $k_{contact}$ simulation has excellent agreement with measurement at the outset of the deposition. However, the simulation overpredicts temperatures when diverges from the experiment, which indicates that the contact pressure along the perimeter increases during deposition. This may occur due to longitudinal distortion, angular distortion, or both.

This analysis of the temperature history, gap conductance values, and pressure negates two fundamental components of the model. First, Approximation 3, *Gap conductance is constant over the deposition history*, does not appear to be valid for the bench clamped cladding experiment. Second, the assumption underlying decoupled analyses (Assumption 10), which assumes thermal behavior is independent of mechanical behavior, is not valid for large contact area cladding simulations. To attain results with the same accuracy of the cantilevered simulation, modifications to the model are necessary. Possibly a fully coupled approach could capture the entirety of the measured temperature histories. Alternatively, implementing time and location dependent gap conductance may allow further calibration of gap conductance over the course of the simulation. Even within the limitations of the decoupled modeling approach, significant improvement in measurement–model correlation is attained once gap conductance elements have been implemented. Below, a quantified comparison of modeled thermal losses for the two experimental setups further illustrates the significance of conduction losses in DED processes.

4.5.3 Thermal Loss Modes

Table 4.6 shows the percentage of simulated losses due to convection, radiation, and conduction for both the cantilevered and bench cases. The cantilevered case, with a contact area of 2.5% has conduction losses around 2%, indicating contacting bodies may be ignored in cases of minimal contact area for basic analyses. Bench based cladding, with a contact area of 40%, loses roughly 70–80% of the absorbed heat via conduction. For this and similar workbench mounted cladding, conduction losses must be accounted for in the model to produce reasonably accurate thermal results.

Comparative losses due to conduction versus due to the combined surface losses of conduction and radiation are depicted in Figure 4.14 using two arbitrary gap conductances and a combined heat transfer coefficient of 20 W/m^2/°C. For a high contact pressure and thus gap conductance, any amount of contact produces significant conduction losses. For the lower gap conductance, corresponding to a case with low contact pressure, conductive losses do

TABLE 4.6   Percentage of model losses by each heat transfer mode

Thermal loss mode	Cantilevered 1% $k_{contact}$	Bench 5% $k_{contact}$	Bench 1% $k_{contact}$
Convection (%)	56	11	21
Radiation (%)	41	5	9
Conduction (%)	2	84	70

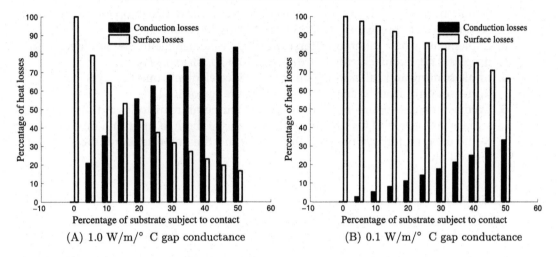

(A) 1.0 W/m/° C gap conductance   (B) 0.1 W/m/° C gap conductance

FIGURE 4.14   Comparison of relative conduction and surface losses.

not become significant until 10–20% of the total surface area is in contact. Thus for most industrial practices where parts are bench-bolted, even when the contact pressure is relatively low, conduction losses must be accounted for in the model.

4.6 CONCLUSIONS

The impact of fixturing in the thermal modeling of laser cladding has been investigated using a validated finite element model. Examination of both measured and modeled temperature histories for two cladding experimental setups, one using a cantilevered fixture with minimal contact area and one fixed to a work bench with a high contact area, shows the significance of conduction losses during laser cladding. It has been shown that accounting for fixturing in DED FE thermal simulations improves thermal model accuracy, which will improve the accuracy of any subsequent mechanical or microstructural model. A strategy for calibrating gap conductance from the theoretical maximum value and implementing gap conductance elements into the FE modeling of laser cladding has been presented. Improvements in simulation accuracy using gap conductance elements for the cantilevered fixtured model (with 2% contact area) was limited to the near deposition region. The fixtured work bench model (with 40% contact area) exhibited a more significant increase in simulation–

experiment correlation when using calibrated gap conductance, decreasing the percent error by at least 37% at each of the four thermocouples. A brief summary of the conclusions is presented below.

1. Conduction losses near the area of contact for cantilevered fixtured components have shown to be significant, but may not impact the thermal history further away from the fixture.
2. Conduction losses have been shown to be significant for benchtop fixtured laser cladding, accounting for approximately 70–80% of the total losses.
3. Implementing and calibrating gap conductance elements into the cantilevered simulation decreases near-contact temperature error from 20.5% to 6.49%.
4. Implementing and calibrating gap conductance elements into the work bench simulation decreases error from a range of 60–102% to a range of 10–45%.
5. Comparison of experimental and simulated temperature histories indicate that the contact surface pressure, and therefore gap conductance, is both location and time dependent, due to the unequal force applied by the fixture and the deformation of the substrate during deposition.

References

[1] Griffith M, Keicher D, Atwood C, Romero J, Smugeresky J, Harwell L, et al. Free form fabrication of metallic components using laser engineered net shaping (LENS). In: P. solid freeform fab. symp.s, vol. 9. The University of Texas at Austin; 1996. p. 125–31.
[2] Mazumder J, Choi J, Nagarathnam K, Koch J, Hetzner D. The direct metal deposition of H13 tool steel for 3-D components. JOM-J Min Met Mat S 1997;49(5):55–60.
[3] Griffith M, Schlienger M, Harwell L, Oliver M, Baldwin M, Ensz M, et al. Understanding thermal behavior in the LENS process. Mater Des 1999;20(2):107–13.
[4] Heigel J, Michaleris P, Palmer T. In situ monitoring and characterization of distortion during laser cladding of Inconel 625. J Mater Process Technol 2015;220:135–45.
[5] Gouge M, Heigel J, Michaleris P, Palmer T. Modeling forced convection in the thermal simulation of laser cladding processes. Int J Adv Manuf Technol 2015;79(1–4):307–20.
[6] Cooper M, Mikic B, Yovanovich M. Thermal contact conductance. Int J Heat Mass Transf 1969;12(3):279–300.
[7] Zavarise G, Wriggers P, Stein E, Schrefler B. Real contact mechanisms and finite element formulation—a coupled thermomechanical approach. Int J Numer Methods Eng 1992;35(4):767–85.
[8] Song S, Yovanovich M, Goodman F. Thermal gap conductance of conforming surfaces in contact. J Heat Transf 1993;115(3):533–40.
[9] Nishino K, Yamashita S, Torii K. Thermal contact conductance under low applied load in a vacuum environment. Exp Therm Fluid Sci 1995;10(2):258–71.
[10] Wahid S, Madhusudana C. Thermal contact conductance: effect of overloading and load cycling. Int J Heat Mass Transf 2003;46(21):4139–43.
[11] Gmelin E, Asen-Palmer M, Reuther M, Villar R. Thermal boundary resistance of mechanical contacts between solids at sub-ambient temperatures. J Phys D, Appl Phys 1999;32(6):R19.
[12] Kar A, Mazumder J. One-dimensional diffusion model for extended solid solution in laser cladding. J Appl Phys 1987;61(7):2645–55.
[13] Ghosh S, Choi J. Modeling and experimental verification of transient/residual stresses and microstructure formation in multi-layer laser aided DMD process. Trans-Am Soc Mech Eng J Heat Trans 2006;128(7):662.
[14] Anca A, Fachinotti V, Escobar-Palafox G, Cardona A. Computational modelling of shaped metal deposition. Int J Numer Methods Eng 2011;85(1):84–106.
[15] Zhu G, Zhang A, Li D, Tang Y, Tong Z, Lu Q. Numerical simulation of thermal behavior during laser direct metal deposition. Int J Adv Manuf Technol 2011;55(9–12):945–54.

[16] Hao M, Sun Y. A FEM model for simulating temperature field in coaxial laser cladding of TI6AL4V alloy using an inverse modeling approach. Int J Heat Mass Transf 2013;64:352–60.

[17] Tseng W, Aoh J. Simulation study on laser cladding on preplaced powder layer with a tailored laser heat source. Opt Laser Technol 2013;48:141–52.

[18] Hoadley A, Rappaz M. A thermal model of laser cladding by powder injection. Metall Trans B 1992;23(5):631–42.

[19] Hofmeister W, Wert M, Smugeresky J, Philliber J, Griffith M, Ensz M. Investigation of solidification in the laser engineered net shaping process. J Minerals Metals Mater Soc 1999;51(7):1–6.

[20] Klingbeil N, Beuth J, Chin R, Amon C. Residual stress-induced warping in direct metal solid freeform fabrication. Int J Mech Sci 2002;44(1):57–77.

[21] Aggarangsi P, Beuth J, Gill D. Transient changes in melt pool size in laser additive manufacturing processes. In: Solid freeform fabrication proceedings. Solid freeform fabrication symposium. Austin, TX: University of Texas; 2004. p. 163–74, in print.

[22] Wang L, Felicelli S. Analysis of thermal phenomena in LENS™ deposition. Mater Sci Eng A 2006;435:625–31.

[23] Wang L, Felicelli S. Process modeling in laser deposition of multilayer SS410 steel. J Manuf Sci Eng 2007;129(6):1028–34.

[24] Kamara A, Marimuthu S, Li L. A numerical investigation into residual stress characteristics in laser deposited multiple layer waspaloy parts. J Manuf Sci Eng 2011;133(3):031013.

[25] Paul R, Anand S, Gerner F. Effect of thermal deformation on part errors in metal powder based additive manufacturing processes. J Manuf Sci Eng 2014;136(3):031009.

[26] Michaleris P, DeBiccari A. Prediction of welding distortion. Weld J-Including Weld Res Suppl 1997;76(4):172.

[27] Ageorges C, Ye L, Mai Y, Hou M. Characteristics of resistance welding of lap shear coupons. Part I: Heat transfer. Composites, Part A, Appl Sci Manuf 1998;29(8):899–909.

[28] Khandkar M, Khan J, Reynolds A. Prediction of temperature distribution and thermal history during friction stir welding: input torque based model. Sci Technol Weld Join 2003;8(3):165–74.

[29] Soundararajan V, Zekovic S, Kovacevic R. Thermo-mechanical model with adaptive boundary conditions for friction stir welding of Al 6061. Int J Mach Tools Manuf 2005;45(14):1577–87.

[30] Colegrove P, Shercliff H, Zettler R. Model for predicting heat generation and temperature in friction stir welding from the material properties. Sci Technol Weld Join 2007;12(4):284–97.

[31] Li T, Shi Q, Li HK. Residual stresses simulation for friction stir welded joint. Sci Technol Weld Join 2007;12(8):664–70.

[32] Awang M, Mucino V. Energy generation during friction stir spot welding (FSSW) of Al 6061-T6 plates. Mater Manuf Process 2010;25(1–3):167–74.

[33] Hamilton C, Dymek S, Sommers A. A thermal model of friction stir welding in aluminum alloys. Int J Mach Tools Manuf 2008;48(10):1120–30.

[34] Zain-ul Abdein M, Nelias D, Jullien J, Deloison D. Prediction of laser beam welding-induced distortions and residual stresses by numerical simulation for aeronautic application. J Mater Process Technol 2009;209(6):2907–17.

[35] Yu M, Li W, Li J, Chao Y. Modelling of entire friction stir welding process by explicit finite element method. Mater Sci Technol 2012;28(7):812–7.

[36] Wang H, Colegrove P, Mehnen J. Hybrid modelling of the contact gap conductance heat transfer in welding process. Adv Eng Softw 2014;68:19–24.

[37] Li H, Simplified Liu D. Thermo-mechanical modeling of friction stir welding with a sequential FE method. Int J Model Optim 2014;4(5):410.

[38] Mughal M, Fawad H, Mufti R. Three-dimensional finite-element modelling of deformation in weld-based rapid prototyping. J Mech Eng Sci 2006;220(6):875–85.

[39] Floreen S, Fuchs G, Yang W. The metallurgy of alloy 625. Superalloys 1994;718(625):13–37.

[40] Denlinger E, Irwin J, Michaleris P. Thermomechanical modeling of additive manufacturing large parts. J Manuf Sci Eng 2014;136(6):061007.

[41] Michaleris P. Modeling metal deposition in heat transfer analyses of additive manufacturing processes. Finite Elem Anal Des 2014;86:51–60.

5

Microstructure and Mechanical Properties of AM Builds

Allison M. Beese

The Pennsylvania State University, USA

5.1 INTRODUCTION

As deformation and fracture initiate on a material's microscale, the microstructure of metals, namely the grain size and morphology, texture, and phases present, impacts the mechanical properties of metallic materials. In additive manufacturing (AM) of metals, due to the layer-by-layer deposition process, the material is subjected to complex thermal histories that differ from those seen in casting, welding, or conventionally heat treated materials. In particular, metal feedstock is delivered to a location, melted, and it then rapidly solidifies and fuses to the layer below, and is reheated and cooled as additional material is deposited next to and above each deposited bead of material.

In AM, solidification rates can reach 1000–6000 K/sec [1–3], which is orders of magnitude higher than the solidification rates seen in casting of 0.1–80 K/sec [4,5]. Additionally, while the cooling rates seen in AM may be on a similar order as those seen in a variety of welding techniques, the repeated thermal cycles as layers are continually added in AM are not present in welding.

The rapid solidification and complex thermal histories in metal AM impact the microstructures (grain sizes and phases present) of the deposited materials, which in turn dictate the mechanical properties (including hardness, yield strength, ultimate tensile strength, ductility, and fatigue properties) of the components. Due to the complex thermal history, the microstructure is typically anisotropic and heterogeneous, which has the potential to result in anisotropic, heterogeneous material properties within a component. This chapter discusses the microstructure and mechanical properties in additively manufactured metals.

Thermo-Mechanical Modeling of Additive Manufacturing
DOI: 10.1016/B978-0-12-811820-7.00007-0

5.2 EXPERIMENTAL CHARACTERIZATION

5.2.1 Microstructure

Microstructures in components made by AM, namely the grain size and morphology as well as texture and microstructural phases present, are characterized using a variety of different methods. Optical and scanning electron microscopy (OM and SEM) can be used to quantify grain size and morphology, where slices in at least three planes are required for full information of the 3-dimensional grain morphology in AM. X-ray diffraction can be used to determine bulk texture of the sample. Electron backscattered diffraction (EBSD) can be used to determine local texture, as well as grain morphology in samples whose grain structure is not visible by OM or SEM.

To determine phases present, X-ray diffraction can be used to probe volumes on the surface in the order of mm in diameter by μm in depth, identifying phases that constitute more than approximately 2–5 vol% of the probed area [6], and providing bulk information on phases present. For finer scales, EBSD inside an SEM can be used as it distinguishes microstructural phases by their crystal structure. Energy dispersive spectroscopy (EDS) in an SEM can be used in concert with EBSD to quantify the elemental composition of phases in order to confirm EBSD phase identification. For nanoscale measurements, transmission electron microscopy (TEM) can be used to both measure the elemental composition via EDS, as well as to measure the crystal lattice parameters using selected area electron diffraction (SAED).

5.2.2 Hardness

Hardness testing is performed to characterize the resistance of the material to plastic deformation [7]. It is related to the yield strength of the material, and is a relatively non-destructive method for assessing mechanical properties as it results in a surface indent in the order of 10s of μm or less in each dimension. For hardness characterization, a rigid indenter is pressed into the sample, and generally, the force used to make the indent is normalized by the size of the imprint, to compute hardness [8]. Vickers microhardness is frequently used in assessing the hardness of additively manufactured materials, as it allows for high spatial resolution (measurements can be taken on the order of every 200 μm with an indent size of 50 μm). Nanoindentation may also be used for finer measurements.

5.2.3 Yield and Ultimate Tensile Strengths, Elongation

Uniaxial tension tests are the most commonly used destructive test method for evaluating the mechanical properties of additively manufactured materials. By assuring that samples comply with ASTM E8 [9], a tensile test can be used to measure the elastic modulus, yield strength, ultimate tensile strength, and elongation of the material.

5.2.4 Fatigue

For the adoption of additive manufacturing in structural components, fatigue, or the material behavior under repeated cyclic loading, must be quantified and repeatable. The most

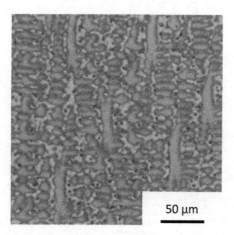

50 µm

FIGURE 5.1 Optical micrograph of microstructure of Inconel 625 fabricated by gas tungsten arc welding additive manufacturing showing dendrites extending in the build (vertical) direction. Figure from [17].

commonly used experiments to evaluate fatigue performance are fatigue crack growth and fatigue life experiments. Fatigue crack growth tests are performed on pre-notched specimens with a sharp crack in order to determine the rate of crack growth for given loading conditions [10]. Fatigue life tests are performed on smooth samples at varying axial or bending alternating stress amplitudes to determine the number of cycles a material can withstand before failure [11,12].

5.3 EXPERIMENTAL RESULTS

As the microstructure in a material dictates its mechanical properties, this section will discuss first the microstructure in additively manufactured metals and how these connect to the resulting properties.

5.3.1 Microstructure

5.3.1.1 Inconel 625

Inconel 625 is a solid solution strengthened Ni-base superalloy, which has high strength at high temperatures [13]. The microstructure of conventionally processed Inconel 625 consists of equiaxed grains of the face-centered cubic (fcc) gamma phase, with small amounts of Laves phase and carbides possible [14–16]. During powder bed fusion or directed energy deposition of Inconel 625, dendrites form along the build direction, which is the direction of the highest temperature gradient during deposition (see Figure 5.1), and elongated grains form in this direction as well [17]. Additionally, the microstructure in additively manufactured Inconel 625 is highly textured [18,19].

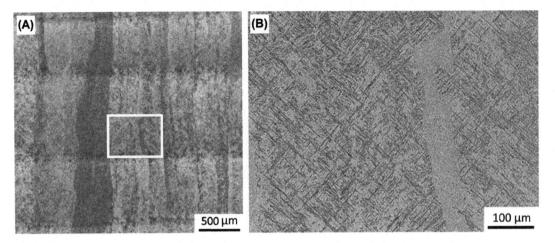

FIGURE 5.2 Optical micrographs of Ti-6Al-4V fabricated by directed energy deposition showing (a) prior beta grains extending in the vertical (build) direction and crossing multiple horizontal layer bands, and (b) fine alpha structure within prior beta grains (inset from (a)). Figure from [31].

5.3.1.2 Ti-6Al-4V

Ti-6Al-4V consists of a V-stabilized body-centered cubic (bcc) beta phase and an Al-stabilized hexagonal close-packed (hcp) alpha phase at room temperature. The microstructure in conventionally processed Ti-6Al-4V varies drastically depending on thermal processing history; the microstructure of as-cast Ti-6Al-4V contains colonies of lamellar alpha, while that of annealed Ti-6Al-4V contains equiaxed grains with coarsened alpha lamellae [20].

Additively manufactured Ti-6Al-4V tends to have a fine Widmanstätten, or basketweave, microstructure, contained within large prior-beta grains that extend multiple deposition layers (see Figure 5.2) [21–31]. Additionally, for very fast cooling rates, typically seen with localized heat sources, acicular or alpha-prime martensite may form [21,24,31].

5.3.1.3 Austenitic Stainless Steel

The microstructure of conventionally processed austenitic stainless steels contains primarily equiaxed austenite grains, and potentially ferrite stringers [32,33]. Due to the high solidification rates in additive manufacturing, the microstructures seen in austenitic stainless steels deposited by AM are similar to that seen in welding, and contain ferrite dendrites, residual dendrite cores, or a cellular structure, within an austenitic matrix [34–37]. The grains are typically elongated in the build direction (see Figure 5.3).

5.3.2 Hardness

5.3.2.1 Inconel 625

The Vickers hardness in additively manufactured Inconel 625 is higher than that in conventionally-processed. The range of hardness values reported for as-deposited AM Inconel 625 is 224–270 HV [17,38] compared to the maximum hardness values seen in cast or

FIGURE 5.3 Optical image of microstructure of stainless steel 304L fabricated by directed energy deposition showing elongated grains in the build (vertical) direction, and layer bands in the horizontal direction. Figure from [36].

annealed Inconel 625 of 266 HV and 145 HV, respectively [13,39]. The high hardness may be attributed to the presence of secondary phases, as well as fine dendritic structures within grains, in the additively manufactured material, and that are absent in annealed Inconel 625.

5.3.2.2 *Ti-6Al-4V*

The range of Vickers hardness varies drastically in additively manufactured Ti-6Al-4V, even within the same processing method. The range of hardness values reported for Ti-6Al-4V made by wire-based DED AM is 341–355 HV [40] and for electron beam PBF is 368–372 [27]. These are higher than the cast and annealed hardness values for Ti-6Al-4V of 200 HV and 202 HV, respectively [41,42]. This high hardness in additively manufactured Ti-6Al-4V is likely due to the fine alpha lath or martensitic alpha structure, which impedes dislocation motion, compared to relatively wider laths of alpha in conventionally processed Ti-6Al-4V.

5.3.2.3 *Austenitic Stainless Steel*

The Vickers hardness in additively manufactured AISI type 316/L and 304/L austenitic stainless steels is generally higher than that in conventionally-processed austenitic stainless steels. In particular, in stainless steel 316L, the range of hardness values reported for DED AM is 164–350 HV [43–47] and for PBF is 132–235 HV [37,48], compared to that of annealed 316L, which ranges from 215 to 225 HV [46].

In 304L stainless steel deposited by PBF AM, the hardness has been reported to range from 209 to 217 HV [49] compared to the hardness in annealed 304L of 136 HV [50]. The disparity in hardness is likely due to the fine microstructural features in additively manufactured austenitic stainless steels, which include finely spaced ferrite dendrites or dendritic cores, cellular structures, and may also be impacted by the high dislocation density and to some level, residual stresses, present in the additively manufactured parts, which are reduced or eliminated in conventional annealing.

5.3.3 Yield Strength, Ultimate Tensile Strength, and Elongation

In general, the yield and ultimate tensile strengths in as-deposited metallic alloys fabricated by additive manufacturing are higher than those found in their conventionally processed counterparts due to fine microstructural features in parts made by AM resulting from the rapid solidification during processing. However, the elongation, or ductility, in metals made by AM is typically less than that found in their conventionally processed counterparts. This is due to a combination of fine microstructural features impeding dislocation motion, leading to dislocation pileup, as well as round and sharp porosity in the components made by AM, which reduces ductility.

5.3.3.1 Inconel 625

The majority of the literature on Inconel 625 is on its properties when deposited as a clad layer to strengthen or repair the outer layer of a component. However, there is some literature on the mechanical properties of 3D additively manufactured Inconel 625.

The yield and ultimate tensile strengths in additively manufactured Inconel 625 span the properties found in conventionally annealed Inconel 625. For Inconel 625 made by AM, the range of yield strength values reported in the literature is 330–1070 MPa, and the range of ultimate tensile strengths is 684–1030 MPa [17,38,51]. Conventionally annealed Inconel 625 has a yield strength of 430 MPa and ultimate tensile strength of 930 MPa [21].

The ductility of additively manufactured Inconel 625 varies widely above and below that reported for conventionally annealed Inconel 625, with tensile elongations in AM Inconel 625 ranging from 8 to 69% [17,38,51] compared to approximately 52% in conventionally annealed Inconel 625 [21].

5.3.3.2 Ti-6Al-4V

In general, the yield and ultimate tensile strengths are higher and ductility lower in additively manufactured Ti-6Al-4V compared to its annealed or cast counterparts; however, a range of properties are reported in the literature. The range of yield strength values reported for DED AM Ti-6Al-4V, with wire or powder feedstock, is 522–1105 MPa [25,30,40, 52–54] and for PBF is 736–1330 MPa [22,24,26–28,54–56]. Typical values of yield strengths for cast and annealed Ti-6Al-4V are 896 MPa and 855 MPa, respectively [20,41].

The range of ultimate tensile strength values reported for DED AM Ti-6Al-4V, with wire or powder feedstock, is 797–1163 MPa [25,30,40,52–54] and for PBF, with a laser or electron beam heat source, is 967–1400 MPa [22,24,26–28,54,55]. Typical values of ultimate tensile strengths for cast and annealed Ti-6Al-4V are 1000 MPa and 930 MPa, respectively [20,41].

The ductility of additively manufactured Ti-6Al-4V is generally lower than the ductility of annealed, and lower than or equal to that of cast, Ti-6Al-4V, which have tensile elongations of 12% and 8%, respectively [20,41]. The ductility in DED AM Ti-6Al-4V has been reported to range in 1–19% [25,30,40,54,57,58] with that in PBF AM Ti-6Al-4V ranging in 2–15% [22,24,26–28,54,59–62]. The wide range of ductility values seen in additively manufactured Ti-6Al-4V can be largely attributed to internal defects, such as sharp lack-of-fusion

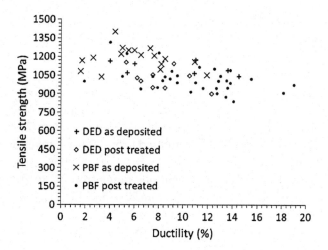

FIGURE 5.4 Tensile strength versus ductility for Ti-6Al-4V fabricated by directed energy deposition and powder bed fusion in as-deposited and heat treated conditions. Figure from [31].

pores between layers, which act as stress risers under load, reducing ductility [31]. Porosity is challenging to assess and quantify; therefore, it is frequently not reported in the literature. However, fractography studies may be used to determine if preexisting porosity was the cause of low ductility [56,63].

While in general the tensile strength is found to decrease with ductility in Ti-6Al-4V made by different AM methods, there is significant scatter in this data, as shown in Figure 5.4. This figure also highlights the ability to increase ductility slightly with heat treatments, with more details provided in [31].

5.3.3.3 Austenitic Stainless Steel

The yield strengths in additively manufactured austenitic stainless steels are typically greater than those in their conventionally-processed counterparts. The most commonly studied austenitic stainless steels made by AM are AISI type 316/316L and 304/304L.

The range of yield strength values reported for DED AM 316/316L stainless steel is 274–593 MPa [43–47,64–66] and for PBF is 287–640 MPa [37,48,59,67–70]. These are higher than for conventional 316L in the cast or annealed conditions, whose yield strength ranges from 241 to 365 MPa [46,67].

The range of yield strength values reported for DED AM 304/304L stainless steel is 274–448 MPa [36,65] and for PBF is 156–520 MPa [49,71]. These are generally higher than for conventional 304L in the annealed condition, whose yield strength ranges in 168–265 MPa [36,50].

The ultimate tensile strengths in additively manufactured austenitic stainless steels vary significantly between studies. The range of ultimate strengths reported for DED AM 316/316L stainless steel is 430–970 MPa [43–47,64–66] and for PBF is 501–760 MPa [37,48,59, 67–70]. These encompass the ultimate tensile strengths of conventional 316L in the cast or annealed conditions, which range in 586–596 MPa [46,67].

The ultimate tensile strengths in DED AM 304/304L stainless steel are 560–710 MPa [36,65] and for PBF are 389–710 MPa [49,71]. These are within the range of conventional 304L in the annealed condition, whose ultimate tensile strength ranges in 556–722 MPa [36,50].

The ductility of additively manufactured 316/316L stainless steel is generally more scattered and lower than that in conventionally processed 316/316L stainless steel. The tensile elongations reported in literature for AM 316/316L stainless steel range from 9 to 70% in 316/316L stainless steel made by DED [43–47,64–66] and in 7–42% in 316/316L stainless steel made by PBF [37,48,59,67–70] compared to 50–70% in conventionally annealed 316/316L stainless steel [46].

In DED 304/304L, the tensile elongations range from 42 to 70% [36,65], and in PBF AM 304/304L, the tensile elongations range from 22 to 58% [49,71]. Conventionally annealed 304L has tensile elongations between 61 and 63% [36,50].

5.3.4 Fatigue

5.3.4.1 Ti-6Al-4V

The fatigue properties of additively manufactured Ti-6Al-4V have been investigated primarily because of its potential to be used in custom biomedical implants or cyclically loaded components in aerospace applications. A thorough review of the impact of post-processing (heat treatment and machining) on the fatigue behavior of Ti-6Al-4V made by AM is presented in [72]. This review focused on comparison of fatigue life curves for different AM methods, with and without heat treat, and with and without surface machining. In general, it was shown that heat treating increased the fatigue life of samples, which can be attributed to the coarsening of microstructural features (e.g., α-lath widths and prior-β grains) as well as the relieving of residual stresses incorporated during deposition. Additionally, as the rough surface of a sample can provide nucleation sites for fatigue cracks, it was found that in every situation examined (powder bed fusion with a laser heat source, electron beam melting, directed energy deposition with powder), machining of the surface, to decrease surface roughness, resulted in significant increases in fatigue life. However, it is noted that significant scatter is present in the data, which may be partially attributed to variations in processing conditions (e.g., laser power, laser scanning speed) as well as internal porosity (e.g., pore volume fraction, pore size distribution, pore shape distribution).

A study on fatigue crack growth by Leuders and co-workers focused on the impact of heat treatment as well as hot isostatic pressing (HIP) on the fatigue threshold and the rate of crack growth [60]. This team found that both heat treatment, which coarsened microstructural features and relieved residual stresses, and hot isostatic pressing, which ideally closed internal pores, coarsened microstructures, and relieved residual stresses, resulted in similar improvements in the fatigue resistance of additively manufactured Ti-6Al-4V. They therefore concluded that the primary need for heat treatment was to coarsen grains and reduce residual stresses, rather than to close all internal pores.

5.4 DISCUSSION

A major challenge in linking processing, structure, and mechanical properties in additively manufactured metallic components is the sparseness of information on the processing parameters used, as well as often minimal reporting on microstructure. In addition to the fact that there are numerous commercial and custom-built AM machines that researchers are using to deposit parts, even within a single system, there is variation.

In particular, since numerous parameters can be varied to create the same part (e.g., AM method, laser power, scanning speed, layer height, hatch spacing, time between layers, scan strategy, preheat, powder and gas flow rate for DED), it is challenging to compare data from disparate studies.

Variations in ductility are common in additively manufactured components, owing to the potential anisotropy in ductility as well as the presence of internal porosity. Lack-of-fusion porosity, in which a layer in AM does not completely melt and infiltrate the layer below, is often long and flat, and oriented between single beads of deposited material. Therefore, this type of porosity negatively impacts the ductility in the build direction, as the sharp edges act as stress risers, while these pores have less of an impact on ductility perpendicular to the build direction [31]. Therefore, within a single study, or when comparing data from multiple studies, the possible presence of internal porosity impacts the reported ductility as well as the statistical variation in ductility among studies.

5.5 CONCLUSIONS

There is an ongoing need for theoretical, computational, and experimental studies to quantitatively link processing, microstructure, and mechanical properties in parts made by AM. A major challenge in interpreting results in the literature is that many of the processing conditions are not reported (processing parameters: laser power, scanning speed, hatch spacing, laser spot size, layer height, scan strategy, gas and powder flow rate in DED), or are not measured (e.g., powder capture efficiency in DED, spatial thermal history in all AM methods). *In situ* diagnostics of the AM process, including thermal imaging of the melt pool and measurement of the full spatial thermal history, will aid in the development of quantitative links between processing and microstructure, while *ex situ* studies are needed for quantitative linking of structure to properties.

References

[1] Rai R, Elmer JW, Palmer TA, DebRoy T. Heat transfer and fluid flow during keyhole mode laser welding of tantalum, Ti–6Al–4V, 304L stainless steel and vanadium. J Phys D, Appl Phys 2007;40(18):5753–66.
[2] Kobryn PA, Semiatin SL. The laser additive manufacture of Ti-6Al-4V. JOM 2001;53(9):40–2.
[3] Manvatkar V, De A, DebRoy T. Spatial variation of melt pool geometry, peak temperature and solidification parameters during laser assisted additive manufacturing process. Mater Sci Technol 2015;31(8):924–30.
[4] Inoue H, Koseki T. Clarification of solidification behaviors in austenitic stainless steels based on welding process. Nippon Steel Tech Rep 2007;95:62–70.
[5] Górny M, Tyrała E. Effect of cooling rate on microstructure and mechanical properties of thin-walled ductile iron castings. J Mater Eng Perform 2013;22(1):300–5.

[6] Jenkins R, Snyder RL. Introduction to X-ray powder diffractometry. New York: John Wiley & Sons, Inc.; 1996.

[7] Tabor D. The hardness of metals. Oxford: Clarendon Press; 1951.

[8] International A. ASTM E140-12be1 standard hardness conversion tables for metals relationship among brinell hardness, vickers hardness, rockwell hardness, superficial hardness, knoop hardness, scleroscope hardness, and leeb hardness; 2012. West Conshohocken, PA.

[9] ASTM standard E8/E8M-15a: standard test methods for tension test of metallic materials. In: ASTM international; 2015.

[10] Astm E2760. Standard test method for creep-fatigue crack growth testing. In: ASTM B. stand; 2010. p. 1–19.

[11] ASTM E466-15. Practice for conducting force controlled constant amplitude axial fatigue tests of metallic materials. Reapproved 2002; 2015. p. 1–6.

[12] ASME. E468-11. Standard practice for presentation of constant amplitude fatigue test results for metallic materials; 2011. p. 1–6.

[13] INCONEL ® alloy 625. Special metals corporation, SMC-020 [Online]. Available http://www.specialmetals.com/assets/documents/alloys/inconel/inconel-alloy-625lcf.pdf, 2006.

[14] Sundararaman M, Mukhopadhyay P, Banerjee S. Carbide precipitation in nickel base superalloys 718 and 625 and their effect on mechanical properties. TMS Superalloys 1997;718:625–706.

[15] Floreen S, Fuchs GE, Yang WJ. The metallurgy of alloy 625. In: Superalloys 718, 625, 706 and various derivatives. The Minerals, Metals & Materials Society; 1994. p. 13–37.

[16] Reed RC. The superalloys: fundamentals and applications. Cambridge: Cambridge University Press; 2006.

[17] Wang JF, Sun QJ, Wang H, Liu JP, Feng JC. Effect of location on microstructure and mechanical properties of additive layer manufactured Inconel 625 using gas tungsten arc welding. Mater Sci Eng A 2016;676:395–405.

[18] Dinda GP, Dasgupta AK, Mazumder J. Laser aided direct metal deposition of Inconel 625 superalloy: microstructural evolution and thermal stability. Mater Sci Eng A 2009;509(1–2):98–104.

[19] Ma D, Stoica AD, Wang Z, Beese AM. Crystallographic texture in an additively manufactured nickel-base superalloy. Mater Sci Eng A 2016;684:47–53.

[20] Donachie MJ. Titanium: a technical guide. 2nd edn. Materials Park, OH: ASM International; 2000.

[21] Shunmugavel M, Polishetty A, Littlefair G. Microstructure and mechanical properties of wrought and additive manufactured Ti-6Al-4V cylindrical bars. Proc Technol 2015;20:231–6.

[22] Edwards P, O'Conner A, Ramulu M. Electron beam additive manufacturing of titanium components: properties and performance. J Manuf Sci Eng Nov. 2013;135(6):061016.

[23] Rafi HK, Starr TL, Stucker BE. A comparison of the tensile, fatigue, and fracture behavior of Ti-6Al-4V and 15-5 PH stainless steel parts made by selective laser melting. Int J Adv Manuf Technol 2013;69(5–8):1299–309.

[24] Zhao X, Li S, Zhang M, Liu Y, Sercombe TB, Wang S, Hao Y, Yang R, Murr LE. Comparison of the microstructures and mechanical properties of Ti–6Al–4V fabricated by selective laser melting and electron beam melting. Mater Des 2016;95:21–31.

[25] Keist JS, Palmer TA. Role of geometry on properties of additively manufactured Ti-6Al-4V structures fabricated using laser based directed energy deposition. Mater Des 2016;106:482–94.

[26] Simonelli M, Tse YY, Tuck C. Effect of the build orientation on the mechanical properties and fracture modes of SLM Ti–6Al–4V. Mater Sci Eng A 2014;616:1–11.

[27] Galarraga H, Lados DA, Dehoff RR, Kirka MM, Nandwana P. Effects of the microstructure and porosity on properties of Ti-6Al-4V ELI alloy fabricated by electron beam melting (EBM). Addit Manuf 2016;10:47–57.

[28] Zhai Y, Galarraga H, Lados DA. Microstructure evolution, tensile properties, and fatigue damage mechanisms in Ti-6Al-4V alloys fabricated by two additive manufacturing techniques. Proc Eng 2015;114:658–66.

[29] Åkerfeldt P, Antti M-L, Pederson R. Influence of microstructure on mechanical properties of laser metal wire-deposited Ti-6Al-4V. Mater Sci Eng A 2016;674:428–37.

[30] Palanivel S, Dutt AK, Faierson EJ, Mishra RS. Spatially dependent properties in a laser additive manufactured Ti–6Al–4V component. Mater Sci Eng A 2016;654:39–52.

[31] Beese AM, Carroll BE. Review of mechanical properties of Ti-6Al-4V made by laser-based additive manufacturing using powder feedstock. JOM 2016;68(3):724–34.

[32] AK steel 304/304L stainless steel product data sheet [Online]. Available: http://www.aksteel.com/pdf/markets_products/stainless/austenitic/304_304l_data_bulletin.pdf, 2013.

[33] AK Steel. Stainless steel 316/316L product data bulletin; 2013. p. 2–4.

[34] Elmer JW, Allen SM, Eagar TW. Microstructural development during solidification of stainless steel alloys. Metall Trans A 1989;20(10):2117–31.

[35] Kou S. Post-solidification phase transformations. In: Welding metallurgy. second edn. New Jersey: John Wiley & Sons, Inc.; 2003. p. 216–9.

[36] Wang Z, Palmer TA, Beese AM. Effect of processing parameters on microstructure and tensile properties of austenitic stainless steel 304L made by directed energy deposition additive manufacturing. Acta Mater 2016;110:226–35.

[37] Sun Z, Tan X, Tor SB, Yeong WY. Selective laser melting of stainless steel 316L with low porosity and high build rates. Mater Des 2016;104:197–204.

[38] Amato K, Hernandez J, Murr LE, Martinez E, Gaytan SM, Shindo PW, Collins S. Comparison of microstructures and properties for a Ni-base superalloy (alloy 625) fabricated by electron beam melting. J Mater Sci Res Mar. 2012;1(2):3–41.

[39] American Casting Company. Materials – alloys poured; 2014.

[40] Brandl E, Baufeld B, Leyens C, Gault R. Additive manufactured Ti-6A1-4V using welding wire: comparison of laser and arc beam deposition and evaluation with respect to aerospace material specifications. Phys Proc 2010;5(PART 2):595–606.

[41] Kobryn PA. Casting of titanium alloys. Wright Lab, Wright-Patterson AFB, OH. 1996. p. 112.

[42] Soares da Rocha S, Luis Adabo G, Elias Pessanha Henriques G, Antônio de Arruda Nobilo M, Luis Adabo G. Vickers hardness of cast commercially pure titanium and Ti-6Al-4V alloy submitted to heat treatments. Braz Dent J 2006;17(2).

[43] Zhang K, Wang S, Liu W, Shang X. Characterization of stainless steel parts by laser metal deposition shaping. Mater Des 2014;55:104–19.

[44] Yu J, Rombouts M, Maes G. Cracking behavior and mechanical properties of austenitic stainless steel parts produced by laser metal deposition. Mater Des 2013;45:228–35.

[45] Li J, Deng D, Hou X, Wang X, Ma G. Microstructure and performance optimisation of stainless steel formed by laser additive manufacturing. Mater Sci Technol 2016:1–8.

[46] Ziętala M, Durejko T, Polański M, Kunce I, Płociński T, Zieliński W, Łazińska M, Stępniowski W, Czujko T, Kurzydłowski KJ, Bojar Z. The microstructure, mechanical properties and corrosion resistance of 316L stainless steel fabricated using laser engineered net shaping. Mater Sci Eng A 2016;677:1–10.

[47] Ma M, Wang Z, Wang D, Zeng X. Control of shape and performance for direct laser fabrication of precision large-scale metal parts with 316L stainless steel. Opt Laser Technol 2013;45(1):209–16.

[48] Tolosa I, Garciandía F, Zubiri F, Zapirain F, Esnaola A. Study of mechanical properties of AISI 316 stainless steel processed by 'selective laser melting', following different manufacturing strategies. Int J Adv Manuf Technol Apr. 2010;51:639–47.

[49] Abd-Elghany K, Bourell DL. Property evaluation of 304L stainless steel fabricated by selective laser melting. Rapid Prototyping J 2012;18(5):420–8.

[50] Qu S, Huang CX, Gao YL, Yang G, Wu SD, Zang QS, Zhang ZF. Tensile and compressive properties of AISI 304L stainless steel subjected to equal channel angular pressing. Mater Sci Eng A 2008;475(1–2):207–16.

[51] Yadroitsev I, Thivillon L, Bertrand P, Smurov I. Strategy of manufacturing components with designed internal structure by selective laser melting of metallic powder. Appl Surf Sci 2007;254(4):980–3.

[52] Dinda GP, Song L, Mazumder J. Fabrication of Ti-6Al-4V scaffolds by direct metal deposition. Metall Mater Trans A 2008;39(12):2914–22.

[53] Alcisto J, Enriquez A, Garcia H, Hinkson S, Steelman T, Silverman E, Valdovino P, Gigerenzer H, Foyos J, Ogren J, Dorey J, Karg K, McDonald T, Es-Said OS. Tensile properties and microstructures of laser-formed Ti-6Al-4V. J Mater Eng Perform 2011;20(2):203–12.

[54] Brandl E, Leyens C, Palm F. Mechanical properties of additive manufactured Ti-6Al-4V using wire and powder based processes. IOP Conf Ser, Mater Sci Eng 2011;26:012004.

[55] Murr LE, Quinones SA, Gaytan SM, Lopez MI, Rodela A, Martinez EY, Hernandez DH, Martinez EY, Medina F, Wicker RB. Microstructure and mechanical behavior of Ti-6Al-4V produced by rapid-layer manufacturing, for biomedical applications. J Mech Behav Biomed Mater 2009;2(1):20–32.

[56] Kasperovich G, Hausmann J. Improvement of fatigue resistance and ductility of TiAl6V4 processed by selective laser melting. J Mater Process Technol 2015;220:202–14.

[57] Carroll BE, Palmer TA, Beese AM. Anisotropic tensile behavior of Ti–6Al–4V components fabricated with directed energy deposition additive manufacturing. Acta Mater 2015;87:309–20.

[58] Zhang XD, Zhang H, Grylls RJ, Lienert TJ, Brice C, Fraser HL, Keicher DM, Schlienger ME. Laser-deposited advanced materials. J Adv Mater 2001;33(1):17–23.

[59] Mertens A, Reginster S, Paydas H, Contrepois Q, Dormal T, Lemaire O, Lecomte-Beckers J. Mechanical properties of alloy Ti-6Al-4V and of stainless steel 316L processed by selective laser melting: influence of out-of-equilibrium microstructures. Powder Metall Jul. 2014;57(3):184–9.

[60] Leuders S, Thöne M, Riemer A, Niendorf T, Tröster T, Richard HA, Maier HJ. On the mechanical behaviour of titanium alloy Ti-Al6-V4 manufactured by selective laser melting: fatigue resistance and crack growth performance. Int J Fatigue 2013;48:300–7.

[61] Vilaro T, Colin C, Bartout JD. As-fabricated and heat-treated microstructures of the Ti-6Al-4V alloy processed by selective laser melting. Metall Mater Trans A 2011;42(10):3190–9.

[62] Ramosoeu MKE, Booysen G, Ngonda TN, Bloemfontein C. Mechanical properties of direct laser sintered Ti-6AL-V4. In: Materials science and technology (MS&T) conference, Columbus, OH, October 16–20, 2011.

[63] Thijs L, Verhaeghe F, Craeghs T, Van Humbeeck J, Kruth JP. A study of the microstructural evolution during selective laser melting of Ti-6Al-4V. Acta Mater 2010;58(9):3303–12.

[64] Griffith ML, Keicher DM, Atwood CL, Romero JA, Smugeresky E, Harwell LD, Greene DL, Smugeresky JE, Harwell LD, Greene DL. Free form fabrication of metallic components using laser engineered net shaping (LENS™). In: Proc. 7th solid free. fabr. symp.; 1996. p. 125–32.

[65] Griffith ML, Ensz MT, Puskar JD, Robino CV, Brooks JA, Philliber JA, Smugeresky JE, Hofmeister WH. Understanding the microstructure and properties of components fabricated by laser engineered net shaping (LENS). MRS proc., vol. 625. 2000.

[66] Xue Y, Pascu A, Horstemeyer MF, Wang L, Wang PT. Microporosity effects on cyclic plasticity and fatigue of LENS™-processed steel. Acta Mater 2010;58(11):4029–38.

[67] Röttger A, Geenen K, Windmann M, Binner F, Theisen W. Comparison of microstructure and mechanical properties of 316L austenitic steel processed by selective laser melting with hot-isostatic pressed and cast material. Mater Sci Eng A 2016;678:365–76.

[68] Mertens A, Reginster S, Contrepois Q, Dormal T, Lemaire O, Lecomte-Beckers J. Microstructures and mechanical properties of stainless steel AISI 316L processed by selective laser melting. Mater Sci Forum May 2014;783–786:898–903.

[69] Zhang B, Dembinski L, Coddet C. The study of the laser parameters and environment variables effect on mechanical properties of high compact parts elaborated by selective laser melting 316L powder. Mater Sci Eng A 2013;584:21–31.

[70] Spierings AB, Starr TL, Wegener K. Fatigue performance of additive manufactured metallic parts. Rapid Prototyping J 2013;19(2):88–94.

[71] Guan K, Wang Z, Gao M, Li X, Zeng X. Effects of processing parameters on tensile properties of selective laser melted 304 stainless steel. Mater Des 2013;50:581–6.

[72] Li P, Warner DH, Fatemi A, Phan N. On the fatigue performance of additively manufactured Ti-6Al-4V to enable rapid qualification for aerospace applications traditionally manufactured Ti-6Al-4V reference data. In: 57th AIAA/ASCE/AHS/ASC struct. struct. dyn. mater. conf.; 2016. p. 1–19.

Understanding Microstructure Evolution During Additive Manufacturing of Metallic Alloys Using Phase-Field Modeling

Yanzhou Ji*, Lei Chen†, Long-Qing Chen*

*The Pennsylvania State University, USA †Mississippi State University, USA

Additive manufacturing (AM), due to its flexibility and capability to deal with parts with complicated geometries, have attracted extensive research attention and is believed to be a promising candidate for accelerating the growth of advanced manufacturing industries. Benefiting from the recent development of manufacturing tools and integrated platforms, AM has been successfully extended to the manufacturing of metallic materials [1–13]. Many industries, such as biomedical and aerospace, are being poised to benefit from the metallic AM. Some examples include (1) on-site, rapid fabrication of metallic bone implants with patient and injury-specific designs, and (2) fabrication of replacement parts in remote locations (e.g. outer space). However, in a technical aspect, a more comprehensive understanding on the "processing-microstructure-properties" correlation is still lacking for the metallic AM. The metallic AM process involves non-uniform temperature distributions and rapid thermal cycles that result in microstructures featured with porosity and anisotropy, which differ drastically from their cast or wrought counterparts. Such different microstructure features critically affect the mechanical properties of the AM builds. Therefore, understanding the microstructure development and evolution during the AM process of metallic alloys is an important prerequisite for the optimization of the AM parameters to achieve desired mechanical properties of the AM builds.

6.1 MICROSTRUCTURES IN ADDITIVELY MANUFACTURED METALLIC ALLOYS

Microstructures are compositional and/or structural inhomogeneities developed during the processing of materials [14]. Microstructure evolution is a kinetic process to reduce the total free energies towards the thermodynamically equilibrium states in a material system under applied external fields [14]. Specifically, for AM of metallic alloys, the complex microstructure evolution characteristics arise from many aspects. On one hand, the multi-component, multi-phase nature of most of the commercial alloys enables the diverse microstructure patterns. The multiple alloying elements added to the alloy system, primarily for the purpose of improving the comprehensive mechanical properties or performances in practical applications, can cause substantial compositional inhomogeneities. Meanwhile, the multiple alloying elements and their inhomogeneous distributions within the system enable the formation of various thermodynamically stable and/or metastable phases. The various possible phase transformation kinetic pathways during the fabrication, treatment and processing of the alloy largely complicate the possible microstructure patterns and evolution paths.

On the other hand, the AM processing conditions further add to the complexity of the microstructures. To densify the initial metallic powders or wires into usable metallic parts, melting or partial melting (e.g., sintering) is required. AM techniques realize this process by a layer-by-layer processing fashion involving multiple rapid cycles. The marked temperature gradient subsequently induces the anisotropic growth behavior of certain microstructure features, resulting in the anisotropic mechanical properties of the builds. Meanwhile, the non-uniform temperature distribution and dynamic temperature variation significantly complicate the phase transformation mechanisms and sequences in terms of both thermodynamics and kinetics principles. Microstructure features rarely reported in conventional manufacturing processes can be observed in the metallic AM. In addition, defects such as voids or pores may develop during the AM process, so augmenting the complexity of the microstructures.

6.1.1 Experimental Observations

Tremendous efforts have been made in experimental characterization of the microstructure features in additively manufactured metallic alloys. Typically, optical microscope (OM), scanning electron microscope (SEM) and transmission electron microscope (TEM) are used for microstructure morphology characterization, and electron backscatter diffraction (EBSD) is used for texture measurements. These experimental observations enable to outline the overall microstructure evolution processes during the metallic AM. The underlying mechanisms of microstructure evolution and their effects on the resulting mechanical properties of the AM builds can be hypothesized and verified as well. Especially, the AM of Ti-6Al-4V (in wt.%) has attracted extensive research interests [11,13,15–31], due to its excellent comprehensive mechanical properties, corrosion resistance and biomedical compatibility. Here an additively manufactured Ti-6Al-4V alloy is used as an example for the illustration purpose.

The AM of Ti-6Al-4V alloys has been accomplished by different AM techniques, and the resulting microstructures are prominently affected by the specific AM technique used. In terms of powder/wire supplies, powder-bed-fusion (PBF) AM techniques [15,17,18,25,30, 32–34] and direct-energy-deposition (DED) techniques [13,16,20,29,31] are both reported for Ti-6Al-4V; in terms of power source, laser beam [11,15,20–22,24,29–35], electron beam [17–19, 25,28,32,36] and arc [26,27] can be applied; in terms of processing methods, melting and sintering (followed by hot isostatic pressing (HIP)) are both reported. Nevertheless, since all of these different AM techniques and their combinations require melting or partial melting to densify the material, the thermal effect plays a critical role in the microstructure development

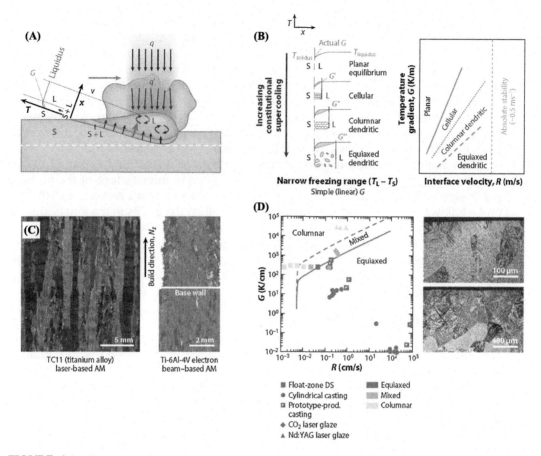

FIGURE 6.1 Illustration of the effect of processing on texture and grains in AM, adapted from [9] (Credit: P.C. Collins et al.)

during the AM building process. For illustration purpose, the microstructure evolution processes during selective electron beam melting (SEBM) PBF of Ti-6Al-4V alloys are described, and the effect of AM techniques on microstructure evolution in Ti-6Al-4V is briefly discussed.

For the building process of each layer in SEBM, micro-scale metal powders of Ti-6Al-4V are first uniformly spread. The electron beam heat source then follows the pre-set scanning path to melt the powders, forming a dynamic melt pool near the scanning probe. The shape of the melt pool depends on not only the specific materials system, but also AM parameters such as the heat source power, the shape of the scanning probe and the scanning speed.

6.1.1.1 Grain Structures and Textures

Figure 6.1 illustrates the grain structure and texture development during AM of Ti-6Al-4V. β grains with body-centered-cubic (bcc) crystal structure develop near the trailing edge of the melt pool, as the melt pool moves, as shown in Figure 6.1A. The β grain growth direction is largely affected by the temperature gradient direction inside the melt pool [9]. Manufactured by specific AM parameters, the β grains may form different dendrite structures, in which columnar grains following the maximum temperature gradient directions are frequently seen (Figure 6.1C); the specific morphology of the β grains depends on the interplay between the temperature gradient G and the interface velocity (or cooling rate) R, as shown in Figure 6.1B. Solute segregation is also observed at the columnar grain boundaries, where the segregation of minor alloying elements may cause the formation of secondary particles. For the melting of the subsequent layer, the pre-existing layers may be partially re-melted; the development of grain microstructures may change if the scanning path changes.

The dynamic non-uniform temperature distribution in the build is one of the primary causes for the development of grain textures in different intersections of the build sample. For example, if the longitudinal section is longer than the wall section, and the scanning probe follows the same zig-zag path on each layer, then the grains in the longitudinal section are mostly columnar along the building direction [37]; whereas in the wall section, smaller grains appear near both walls while larger grains form and develop along the maximum thermal gradient inside the section [17]. These grain textures, as a result of the temperature gradient and cooling rates along the building direction, will cause anisotropies in mechanical properties of the build [6,11,23,36,37].

6.1.1.2 Solid State Phase Transformation

In addition to the grain morphology and textures, the temperature gradient and thermal history during AM will also affect the microstructure evolution inside β grains. When the temperature decreases below the β transus (about $1000\,°C$ for Ti-6Al-4V), the $\beta \rightarrow \alpha$ allotropic phase transformation may take place, forming α products with hexagonal-close-packed (hcp) crystal structure. The specific $\beta \rightarrow \alpha$ transformation modes may differ under different thermal conditions, which do not only contain the competition between diffusional and diffusionless transformation modes, but also, under a fixed transformation mode, involve the interplay between the nucleation and growth of α products, leading to distinct microstructure features. For example, based on experimental observations during continuous cooling of Ti-6Al-4V, the $\beta \rightarrow \alpha$ transformations may take place in different modes under different cooling rates [38–40]:

(1) under small to moderate cooling rates (<20 K/s), the α phase primarily forms near β grain boundaries due to the higher undercooling and heterogeneous nucleation sites, and then develop coarse α colonies consisting of laths of the same α structural variants; with increasing cooling rates, there is a decrease in the α-colony size and an increase in the intra-granular α nucleation sites, resulting in basket-weave-type microstructures;

(2) under large cooling rates, partitionless $\beta \rightarrow \alpha$ transformation, such as massive (20 K/s < R < 410 K/s) and martensitic (fully martensitic when R > 410 K/s) transformation may take place, resulting in acicular α products inside prior β grains.

During AM processes, the situation becomes more intricate since multiple heating-cooling cycles are present, and cooling/heating rates at a selected region are generally not constant. As a result, different α products may sequentially develop and dissolve during thermal cycles, resulting in complicated $(\alpha + \beta)$ two phase mixtures [13]. The size, morphology and textures of α laths and colonies can significantly influence the mechanical properties of the sample [40]. Besides, due to the compositional inhomogeneities originated from solute segregations at either grain boundaries or α/β interface boundaries, minor precipitates or inclusions may also be present inside prior β grains (such as Ti_3Al) or at grain boundaries, which may have different effects on the mechanical properties of the build. Figure 6.1(d) shows the typical $(\alpha + \beta)$ basket-weave microstructures in the AM build of Ti-6Al-4V.

6.1.1.3 *Effect of Different AM Techniques*

As mentioned above, the microstructure morphology may be affected by the different AM techniques. For example, as reviewed by Beese et al. [11], the as-fabricated Ti-6Al-4V by PBF laser melting is generally finer than that by DED laser melting techniques. In as-fabricated Ti-6Al-4V by PBF laser melting, the grain size is generally smaller and the acicular α' martensites are frequently observed, in contrast to the fully laminar α plates in DED-fabricated samples. The primary cause for the microstructure difference is identified to be the different laser spot size for the two techniques. The smaller laser spot size in PBF AM technique leads to smaller melt pools and larger temperature gradient, which result in finer microstructures.

The type of heat source can also affect the microstructure morphology. For example, Ti-6Al-4V builds fabricated by SEBM contain fewer α' martensites (only at top surfaces), in contrast to that fabricated by SLM [2,32]. The microstructure difference can be attributed to the faster moving velocity and the capability for *in situ* heat treatment of SEBM techniques [2].

In addition, the densification methods also influence the final microstructures. For example, comparing SLM with selective laser sintering (SLS), the SLS process generally has lower laser energy input and/or faster scanning speed since the metal powders are only partially melted. The melt pool is generally smaller in SLS and the densification of metal powders is mainly driven by surface tension rather than melting. Since thermal history is also present in SLS, the anisotropic microstructure features, such as columnar grains, are also observed in SLS-fabricated Ti-6Al-4V [41].

Based on these understandings, the microstructure evolution mechanisms can be hypothesized, and the AM parameters that lead to the change of microstructures can be identified. To further understand the processing parameter-microstructure-property relationship, systematic high-throughput experiments should be conducted in the entire AM processing parameter space, which is not only financially expensive but also time-consuming. Meanwhile, due to the difficulties in *in situ* observation of the microstructure evolution process during AM, the direct evidences and details of the microstructure evolution kinetics are usually lost.

6.1.2 Computational Simulations

Recently, with the development of computation tools and numerical methods, the computational simulation has become a promising alternative to the experimental investigations of AM, in a cost-effective manner. However, AM is a complicated process involving the interactions among different applied external fields, resulting in complicated microstructure features. To accurately predict and reconstruct the microstructure evolution process during AM, the realistic AM processing parameters, materials parameters and geometries should be input into the computation models. Moreover, careful validations through quantitative comparison with experimental results should be performed to ensure the robustness of the model. There have been a series of existing attempts on computational simulations of the microstructure evolution during AM.

Since the prominent feature of metal AM is the complicated thermal effects, the initial computational efforts in AM of alloys largely ignored the microstructure aspects. Instead, macroscale prediction of temperature distribution and history as a function of AM parameters such as power, spot size, scanning speed and scanning directions of the heat source by solving the heat equations [28,42–48] were the main computational focus. These simulations can further couple with finite-element-based mechanical models to predict the macroscopic mechanical properties of the AM build. Notably, Michaleris et al. [28,44,45,48] developed a finite-element-based software for thermo-mechanical modeling of AM which can provide accurate solutions with lower mesh density and higher computation efficiency. These efforts, although do not directly predict the microstructure evolution, can provide indications for the geometries of the melt pool, the temperature distributions and thermal history in the build, which lay the foundation for further microstructure predictions.

By coupling with thermal calculations, there have been a few existing efforts to develop computational models for predicting the grain morphology during AM of alloys. For example, Nie et al. [49] used a microscopic stochastic analysis including temperature-dependent nucleation rate, solute diffusion and growth anisotropy, to simulate the dendrite morphology evolution as a function of cooling rate and temperature gradient in IN718. Similar ideas have been applied by Zhou et al. [50] for AM of a stainless steel using Cellular Automata. By assuming that solidification direction is parallel to the local maximum heat flow direction, and the columnar to equiaxed transition occurs beyond a critical G/R (temperature gradient/cooling rate) ratio, Wei et al. [51] applied a simple two-dimensional (2D) grain growth model to predict and validate the grain orientation development during AM of an Al-alloy.

Moreover, the phase-field approach [14,52–61], which uses the diffuse-interface description to avoid the explicit tracking of interfaces, shows the great potential to simulate microstructure evolution during AM of metallic alloys. There have been some initial attempts to employ the phase-field method for modeling microstructure evolution during AM. For example, Gong et al. [32,62] coupled the thermal process modeling with a phase-field model to simulate the dendrite morphology during AM of Ti-6Al-4V, considering the undercooling effect. Lim et al. [63] proposed a preliminary modeling framework using both the phase-field approach and a crystal plasticity finite element (CP-FE) method which could be applicable to microstructure evolution modeling in AM.

However, to date, due to the complexity of the AM process, microstructure evolution modeling of alloys during AM is still at its early stages. Existing investigations mainly focus on the modeling of certain microstructural aspects. The model validations and the construction of an integrated model containing major microstructure features of AM are still lacking. In addition, the temperature distribution and thermal history during the building process may affect the microstructure features at different length scales, ranging from the grain structures developed during solidification with a typical length scale of several microns, to the intragranular microstructures such as precipitations, micro-segregations and defects at nanoscale. Therefore, it is currently numerically challenging and computationally expensive to simulate the cross-length scale microstructure features during AM.

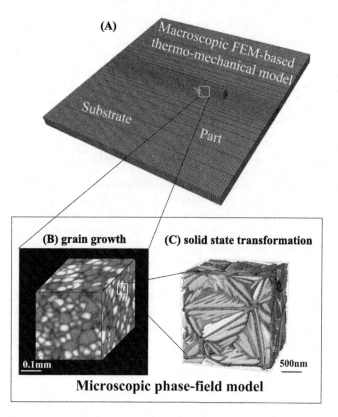

FIGURE 6.2 Illustration of the multi-scale phase-field framework for AM of alloys: (A) finite-element-based thermal model; (B) grain growth phase-field model; (C) sub-grain-scale phase-field model for solid phase transformation.

6.2 MULTI-SCALE PHASE-FIELD MODEL FOR AM OF ALLOYS

In this work, a multi-scale computational framework based on the phase-field approach is presented to simulate the microstructure evolution during AM of alloys. In particular, a SLM or SEBM process is considered. To highlight the major governing factors of the microstructure evolution and simplify the numerical model, the following overall assumptions are made:

(1) The microstructure evolution processes have negligible effect on heat transfer and temperature distribution in the build, while the heat transfer is mainly affected by the heat conductivity and capacity of the material, as well as the AM parameters such as power, scanning probe size and scanning speed of the heat source;

(2) The solidification and the development of grain structures take place in the high temperature regime followed by possible solute segregation and inclusions occur near grain boundaries, while the phase transformation and microstructure evolution inside grains take place in the low temperature regime with negligible grain structure change.

With the assumptions above, the entire microstructure evolution model can be decoupled into three different sub-models on different length scales, as illustrated in Figure 6.2:

(i) the macroscopic thermal model to obtain the temperature distribution and thermal history in the build sample during the whole AM process;

(ii) the grain-scale solidification or grain growth phase-field model to study the grain morphology and texture development and/or solute distribution and segregation;

(iii) the sub-grain-scale phase-field model to simulate the intra-granular phase transformations which may include diffusional transformations such as precipitation and diffusionless structural transformations, depending on the specific materials system.

6.2.1 Linkage Between the Three Sub-Models

Rather than performing the three sets of simulations independently, the linkage among the sub-models is considered to reflect the AM processing conditions. The results of the larger-scale simulations will be used as the input parameters or initial microstructures for the smaller-scale simulations. Specifically, the thermal model focuses on the effect of various AM parameters, such as the power, shape and scanning speed of the heat source, the layer thickness and the scanning paths on the temperature distribution and thermal history of the build. The calculated temperature distribution and thermal history will be input into the grain growth or solidification phase-field model. For simulating the intra-granular microstructure evolution, several representative regions will be selected from the simulated grain structures; the grain structures and solute compositions of the selected region will be enlarged through interpolation methods to obtain better simulation resolutions for the sub-grain microstructures. Since the length scale of the selected region is comparable with the resolution of the thermal model, within the selected region, the temperature distribution can be assumed uniform while changing with time during the AM process. Therefore, thermal history for the selected region will be input into the sub-grain-scale phase-field model to simulate the microstructure evolution. With this multi-scale model, the temperature effect during the AM process on the microstructure evolution can be thoroughly studied. With proper model validation, the effect of the AM parameters on the microstructure evolution of the additively manufactured alloys can also be understood.

6.2.2 Finite-Element Thermal Model

Next, the details of the multi-scale microstructure model for AM of metallic alloys are provided, using Ti-6Al-4V as an example. The model starts from the finite-element-based thermal calculations, using a Life-death Element Technique. Both substrate and printing component are fully discretized into finite elements, but the elements of printing component are inactive before printing. At the beginning of each simulation time step, which corresponds to a real time period during AM, a new group of finite elements is activated to mimic the movement of the heat source, and those active elements that constitute a complete domain are calculated for thermal analysis.

Based on the ABAQUS software platform [64], the temperature field of both the printing component and the substrate can be calculated by solving the 3D heat conduction equation

$$\frac{\partial}{\partial x}\left(k_x\frac{\partial T}{\partial x}\right) + \frac{\partial}{\partial y}\left(k_y\frac{\partial T}{\partial y}\right) + \frac{\partial}{\partial z}\left(k_z\frac{\partial T}{\partial z}\right) + \rho Q = \rho c_p \frac{\partial T}{\partial t} \tag{6.1}$$

(A) **(B)**

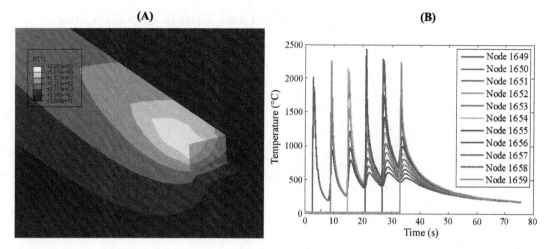

FIGURE 6.3 Calculation results of the thermal model: (A) temperature distribution (in °C) during AM; (B) temperature history of selected RVEs (nodes).

where k_x, k_y and k_z are thermal conductivities along the three coordinate axes, respectively. ρ (kg/m^3) denotes the density, c_p (J/(kg K)) denotes the specific heat capacity, Q (W/kg) denotes the heat source density.

At each simulation step, the convection and radiation boundary conditions are updated and applied to outer surfaces of the active elements

$$k_x \frac{\partial T}{\partial x} n_x + k_y \frac{\partial T}{\partial y} n_y + k_z \frac{\partial T}{\partial z} n_z = h(T_a - T) + \varepsilon_R \sigma_R \left(T_a^4 - T^4 \right) \tag{6.2}$$

where h (W/m^2 K) is the heat convection coefficient, T_a is the ambient temperature, ε_R, σ_R are emissivity and Stefan–Boltzmann constant (5.67×10^{-8} W/m^2 K^4), respectively.

The electron beam heat source is regarded as a body heat flux, which is modeled by Goldak double-ellipsoid model [28]

$$Q = \frac{6\sqrt{3} U I \eta f_s}{a_1 b_1 c_1 \pi \sqrt{\pi}} e^{-\left[\frac{3x^2}{a_1^2} + \frac{3(y + v_Q t)^2}{b_1^2} + \frac{3z^2}{c_1^2} \right]} \tag{6.3}$$

where U represents the electron beam acceleration voltage, I is the electron beam current, η represents the absorption efficiency, f_s represents the process scaling factor, x, y and z are the local coordinates of the heat source, a_1, b_1 and c_1 are the transverse, depth, and longitudinal dimensions of the ellipsoid, respectively, v_Q represents scanning speed of the heat source, and t is the scanning time. The resulting temperature distribution within the build during the AM process and the thermal history in a selected representative volume element (RVE) are shown in Figure 6.3.

6.2.3 Grain-Scale Phase-Field Model: Grain Growth & Solidification

6.2.3.1 Model Description

With the temperature distribution and history at hand, the β grain growth behavior during AM is then investigated using a grain-scale phase-field model. In this model, the texture (crystallographic orientation of each grain) in a simulation cell is specified by a set of continuous order parameters. The total free energy of a polycrystalline microstructure system can be described as follows [65,66]

$$F = \int \left[f_0(\phi_1, \phi_2, ..., \phi_Q) + \sum_{q=1}^{Q} \frac{\kappa_q}{2} (\nabla \phi_q)^2 \right] dr \tag{6.4}$$

where $\{\kappa_q\}$ are positive gradient energy coefficients, and $f_0(\{\phi_q\})$ is the local free energy density, which is defined as

$$f_0(\{\phi_q\}) = -\frac{\alpha}{2} \sum_{q=1}^{Q} \phi_q^2 + \frac{\beta}{4} \left(\sum_{q=1}^{Q} \phi_q^2 \right)^2 + \left(\gamma - \frac{\beta}{2} \right) \sum_{q=1}^{Q} \sum_{s>q}^{Q} \phi_q^2 \phi_s^2 \tag{6.5}$$

in which, α, β and γ are constants, for $\alpha = \beta > 0$ and $\gamma > \beta/2$, f_0 possesses $2Q$ degenerate minima. Those minima are located at $(\phi_1, \phi_2, ..., \phi_Q) = (\pm 1, 0, ..., 0), (0, \pm 1, ..., 0), ...,$ $(0, 0, ..., \pm 1)$, representing the finite number of possible grain orientations in a polycrystal.

Grain growth is described by the temporal and spatial evolution of the order parameters, which yield the time-dependent Ginzburg–Landau equations

$$\frac{\partial \phi_q(\mathbf{r}, t)}{\partial t} = -L_q \frac{\delta F}{\delta \phi_q(\mathbf{r}, t)} \quad (q = 1, 2, ..., Q) \tag{6.6}$$

where $\{L_q\}$ are kinetic rate coefficients influenced by temperature gradient, which can be calculated through the modified Arrhenius type equation such as

$$L_q = A \cdot \left(\frac{T}{T_0} \right)^m \cdot \exp\left(-\frac{E_a}{RT} \right) \tag{6.7}$$

where R, molar gas constant, is 8.314 J/(mol K) and m is a constant lying in the range $-1 < m < 1$, T_0 is the ambient temperature (298.15 K), A is a temperature-independent constant, E_a is the activation energy for interface movement. The evolution equation is solved numerically using finite difference method.

To link the grain-growth phase-field model with the FEM temperature calculations and simulate the grain growth behavior during the layer-by-layer AM process, the following assumptions and approximations are made:

1) When the temperature is above the liquidus temperature (1660 °C), it is set as liquidus temperature;
2) When the temperature falls below the β transus temperature ($\beta \to \alpha$, 1000 °C), the kinetic rate coefficients $\{L_q\}$ are set to be zero.

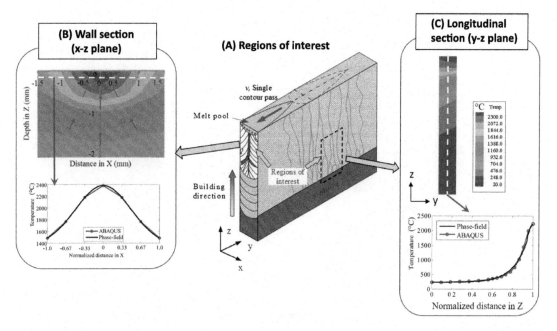

FIGURE 6.4 Connection between the thermal model and the grain growth phase-field model: representative longitudinal section and wall section for grain growth simulation, as well as the interpolated temperature distribution data for phase-field simulations in the two sections.

3) It is assumed that the temperature in the newly added layer is uniform, and the process of applying the uniform temperature is instantaneous;
4) It is assumed that the temperature gradient is fixed after the heat source left, and the direction of maximum heat flux always keeps vertical upward.

With these assumptions, the grain growth behaviors during AM in the longitudinal and wall sections of the build are investigated, due to their distinct temperature distributions. One representative region in each section is selected for the grain growth simulations, as shown in Figure 6.4. To obtain simulated grain morphology with acceptable resolution, the required mesh size for the phase-field simulations ($\sim$μm) should be much less than that of the thermal calculations ($\sim$10 μm). Therefore, the interpolation method is used to fit the temperature distribution data from the thermal calculations, as shown in Figure 6.4, for both longitudinal and wall sections.

6.2.3.2 Simulation Results

With the temperature distribution and history available from the thermal calculations, 2D/3D phase-field grain growth simulations are performed. With randomly distributed grains as the initial state, new layers with adjustable thickness are introduced into the system at adjustable time intervals to mimic the AM process. In the longitudinal section, since the temperature gradient is along the building direction and perpendicular to the layers,

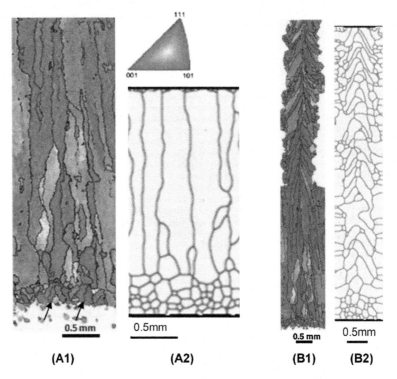

FIGURE 6.5 Comparison between the phase-field simulated grain morphology and experimental observations. (A1) Experimental results in the longitudinal section [37] (credit: A.A. Antonysamy); (A2) simulation results in the longitudinal section; (B1) experimental results in the wall section [17] (credit: A.A. Antonysamy et al.); (B2) simulation results in the wall section.

columnar grains parallel to the building direction are observed (Figure 6.5A2). In the wall section, since the temperature gradient is not uniform as shown in Figure 6.4, grains grow along the local temperature gradient direction to form the columnar grains in the middle and small grains near the edges of the section (Figure 6.5B2). These simulated grain microstructures can well reproduce the experimentally observed ones in both the longitudinal section [37] and the wall section [17], as shown in Figure 6.5, which confirms the effect of temperature distribution during AM on the grain texture development. The simulated grain width, with an average value of 0.21mm, lies within the experimental ranges (0.15-0.3mm).

The development of columnar grains during AM may lead to the anisotropies in mechanical behaviors of the build, which is detrimental in the practical applications. Therefore, the effect on the grain morphology evolution of the AM processing parameters, especially the scanning speed and the layer thickness, is further investigated. Based on the simulations, as briefly illustrated in Figure 6.6, coarse columnar grains develop as the layer thickness and/or the scanning speed decrease, since the larger grains below the newly added layers

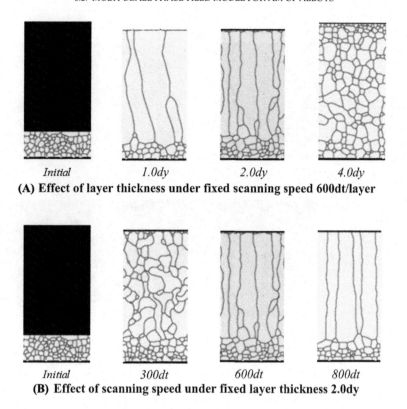

Initial   1.0dy   2.0dy   4.0dy

(A) Effect of layer thickness under fixed scanning speed 600dt/layer

Initial   300dt   600dt   800dt

(B) Effect of scanning speed under fixed layer thickness 2.0dy

FIGURE 6.6   Effect of AM parameters on grain morphology: (A) layer thickness; (B) scanning speed.

have enough time to swallow the smaller grains or select the preferred smaller grains inside the newly added layers, which should be avoided during real AM processes.

6.2.3.3 *Future Directions*

The current phase-field model only focuses on the grain morphology development and ignores the solute distribution during the solidification process. Moreover, even in the longitudinal section, the temperature distribution is not uniform, especially in the melt pool where the solidification and grain growth begins. To more accurately simulate the microstructure evolution in the melt pool and during the movement of the melt pool, the phase-field solidification model by Gong et al. [32,62] should be extended to polycrystal systems and consider the realistic geometry of the melt pool, as illustrated in the preliminary simulation results in Figure 6.7. In addition, as has been experimentally reported, the change of scanning directions of the heat source during AM will also change the grain texture in the longitudinal section; the growth direction of the grains will not exactly follow the maximum temperature gradient direction [67]. These phenomena will be considered in our future phase-field grain growth/solidification models for AM processes.

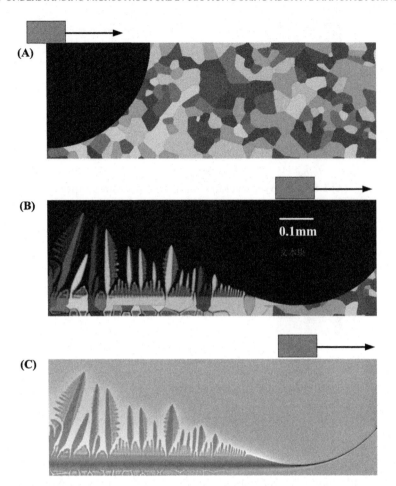

FIGURE 6.7   Preliminary phase-field simulation results for dendrite morphology in a moving melt pool. (A) Initial grain configuration; (B) dendrite morphology; (C) solute distribution. The red rectangle represents the heat source and the arrow represents the scanning direction.

6.2.4 Sub-Grain-Scale Phase-Field Model: Solid-State Phase Transformations

6.2.4.1 Model Description

With the simulated β grain structures, the sub-grain-scale microstructure evolutions are simulated in additively manufactured Ti-6Al-4V alloys. Ignoring the minor inclusions or defects, the $\beta \rightarrow \alpha$ transformation is the major factor for the development of the intra-granular microstructure features. The $\beta \rightarrow \alpha$ transformation happens when the temperature of a local region in the sample is below the β transus temperature ($\sim 1000\,^{\circ}\mathrm{C}$) during AM, or during the follow-up thermo-mechanical processing of the as-built sample. The microstructure features during $\beta \rightarrow \alpha$ transformation is not only affected by the temperature distributions and history, but also influenced by the solute distribution and grain structures. Therefore, different RVEs are selected in the thermal calculation to obtain the local temperature history. The temperature distribution within each RVE is assumed uniform. The grain structures of the corresponding RVE in the grain-scale phase-field simulations are further selected as the initial grain structure. Interpolation method is used to convert the grain structures in the grain growth simulations with larger grid size ($\sim\mu$m) into the sub-grain phase-field simulations with smaller grid size ($\sim$10 nm). Since the $\beta \to \alpha$ transformation may contain both compositional and structural changes, both solute compositions (X_{Al} and X_V) and structure order parameters $\{\eta_{p,g}\}$ are considered to describe the microstructures. Specifically, the Burgers orientation relation for the bcc$\to$hcp transformation is considered, resulting in 12 symmetry-allowed α variants in each β grain [68]. The free energy functional of the system is [69–73]

$$F = \int_V \left(f_{bulk}\left(X_{Al}, X_V, \{\eta_{p,g}\}, T\right) + f_{grad}\left(\{\nabla\eta_{p,g}\}\right) + e_{el} \right) dV \tag{6.8}$$

where f_{bulk} is the temperature-dependent bulk free energy density, f_{grad} is the energy contribution due to the inhomogeneity of order parameters, and e_{el} is the elastic strain energy density.

The bulk free energy density includes the molar Gibbs free energy of all the phases in the systems as well as energy variations due to structure change. Although many stable and metastable phases have been identified in Ti-6Al-4V, the $(\alpha + \beta)$ two phase system is the primary focus of the current study. Therefore, the following bulk free energy density is used:

$$f_{bulk}\left(X_{Al}, X_V, \{\eta_p\}, T\right) = V_m^{-1} \cdot \sum_p h(\eta_p) \cdot f^\alpha\left(X_{Al}^\alpha, X_V^\alpha, T\right)$$

$$+ V_m^{-1} \cdot \left(1 - \sum_p h(\eta_p)\right) \cdot f^\beta\left(X_{Al}^\beta, X_V^\beta, T\right) + g\left(\{\eta_p\}\right) \tag{6.9}$$

where f^α and f^β are molar Gibbs free energies of α and β phases (in J/mol), respectively, which are directly taken from the Ti-Al-V thermodynamic database [74]; V_m is the molar volume, $h(\eta) = 3\eta^2 - 2\eta^3$ is an interpolation function, and $g(\{\eta_p\})$ is a Landau-type double-well potential describing the energy variation due to structure change:

$$g\left(\{\eta_p\}\right) = w \cdot \sum_p \eta_p^2(1 - \eta_p^2) + w' \cdot \sum_{p \neq q} \eta_p^2 \eta_q^2 \tag{6.10}$$

w and w' are parameters characterize the barrier height.

The gradient energy term in Eq. (6.8) is the part of interfacial energy due to the inhomogeneous distribution of order parameters at interfaces. The gradient energy coefficient is assumed to be in the matrix form and the gradient energy is expressed as:

$$f_{grad} = \frac{1}{2} \sum_{p=1}^{12} \kappa_{p,ij}(\nabla_i \eta_p)(\nabla_j \eta_p) \tag{6.11}$$

where $\kappa_{p,ij}$ is the anisotropic gradient energy coefficient tensor for variant p, which, together with the barrier height w, can be obtained from α/β interfacial energy [75] and interface thickness values through thin-interface analysis [76].

The elastic strain energy originates from the mismatch between the crystal structures of both phases. Based on the lattice constants and the lattice correspondence between the crystal structures of the two phases, the stress-free transformation strain (SFTS) can be derived in the local reference coordinates under finite-strain approximation:

$$\varepsilon_{local}^{00} = \begin{pmatrix} \frac{2a_\alpha^2}{3a_\beta^2} - \frac{1}{2} & \frac{a_\alpha^2}{6\sqrt{2}a_\beta^2} & 0 \\ \frac{a_\alpha^2}{6\sqrt{2}a_\beta^2} & \frac{7a_\alpha^2}{12a_\beta^2} - \frac{1}{2} & 0 \\ 0 & 0 & \frac{1}{4}(\frac{c_\alpha^2}{a_\beta^2} - 2) \end{pmatrix} \tag{6.12}$$

where a_α, c_α and a_β are the lattice constants of α and β phases, respectively [77]. The SFTS of each α variant $\varepsilon_{global,ij}^{00}(p)$ can be calculated by axis transformation from the local reference frame to the global reference frame. The elastic strain energy can then be calculated as [78]:

$$e_{el} = \frac{1}{2}C_{ijkl} \cdot (\varepsilon_{ij} - \varepsilon_{ij}^0) \cdot (\varepsilon_{kl} - \varepsilon_{kl}^0) \tag{6.13}$$

where C_{ijkl} is the elastic stiffness tensor, ε_{ij} is the total strain solved from the stress equilibrium equation $\sigma_{ij,j} = 0$ under certain mechanical boundary conditions, ε_{ij}^0 is the overall eigenstrain:

$$\varepsilon_{ij}^0 = \sum_p h(\eta_p) \cdot \varepsilon_{global,ij}^{00}(p) \tag{6.14}$$

The microstructure evolution is governed by the numerical solution of both the Cahn–Hilliard equation (for solute compositions)

$$\frac{\partial X_{Al}}{\partial t} = \nabla \cdot \left(\sum_p h(\eta_p) \cdot \left(\tilde{D}_{AlAl}^\alpha \nabla X_{Al}^\alpha + \tilde{D}_{AlV}^\alpha \nabla X_V^\alpha \right) \right.$$
$$\left. + \left(1 - \sum_p h(\eta_p) \right) \cdot \left(\tilde{D}_{AlAl}^\beta \nabla X_{Al}^\beta + \tilde{D}_{AlV}^\beta \nabla X_V^\beta \right) \right) \tag{6.15a}$$

$$\frac{\partial X_V}{\partial t} = \nabla \cdot \left(\sum_p h(\eta_p) \cdot \left(\tilde{D}_{VAl}^\alpha \nabla X_{Al}^\alpha + \tilde{D}_{VV}^\alpha \nabla X_V^\alpha \right) \right.$$
$$\left. + \left(1 - \sum_p h(\eta_p) \right) \cdot \left(\tilde{D}_{VAl}^\beta \nabla X_{Al}^\beta + \tilde{D}_{VV}^\beta \nabla X_V^\beta \right) \right) \tag{6.15b}$$

where $\tilde{D}_{ij}^\phi$ ($i, j = Al, V; \phi = \alpha, \beta$) are the inter-diffusivity coefficients available from literature [79,80]; and Allen–Cahn equation (for structure order parameters):

$$\frac{\partial \eta_p}{\partial t} = -L \left(\frac{\partial f_{bulk}}{\partial \eta_p} - \kappa_{p,ij} \nabla_i \nabla_j \eta_p + \frac{\partial e_{el}}{\partial \eta_p} \right), \quad p = 1, 2, \ldots, 12 \tag{6.16}$$

where L is a kinetic coefficient related to interface mobility, which can be evaluated according to thin-interface analysis if the interface mobility values are available [76]. To improve the

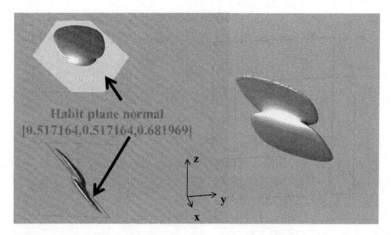

FIGURE 6.8 Morphology of a single α variant in β matrix with the consideration of the anisotropic energy contributions.

computational efficiency, the governing kinetic equations, as well as the stress equilibrium equation are solved using a fast Fourier transform (FFT) algorithm. The stress equilibrium is assumed to be much faster than the microstructure evolution [81].

Due to the anisotropic interfacial energy and SFTS, the morphology of α products, both diffusional and diffusionless ones, are generally anisotropic with lath or acicular shape and habit planes [52,82–86]. With these anisotropies considered, the morphologies of α product are reconstructed, which quantitatively agrees with the experimentally reported $\{3\,3\,4\}$, $\{8\,8\,11\}$, $\{8\,9\,12\}$ or $\{11\,11\,13\}$ habit planes [52,82–84,86], as shown in Figure 6.8, with a maximum deviation of only $3°$. With the energy anisotropies and the temperature-dependent thermodynamic and kinetic coefficients, the non-isothermal growth behavior of α products can be accounted for. To further consider the temperature-dependent nucleation behavior, the classical nucleation theory [59,87,88] is applied:

$$j = Z N_0 \beta^* \exp\left(-\frac{\Delta G^*}{RT}\right) \exp\left(\frac{t}{\tau}\right) \tag{6.17}$$

where j is the nucleation rate, Z is Zeldovich's factor, N_0 is the number of available nucleation sites in the corresponding system (here a simulation cell), β^* is atomic attachment rate, ΔG^* is nucleation barrier, t is elapsed time and τ is incubation time for nucleation. The detailed formulation of these quantities can be found in Kozeschnik et al. [89]. For simplicity, the initial α nuclei are assumed to be spherical. The diffusionless $\beta \rightarrow \alpha$ transformation is assumed to be much faster than the diffusional one [90]. Therefore, the new α nuclei put into the system are assumed to only include order parameter change without composition change.

6.2.4.2 Simulation Results

With the temperature-dependent description of both the nucleation and growth behavior, phase-field simulations are performed in β single crystals under different cooling rates (1 K/s, 10 K/s, 100 K/s). The simulation results shown in Figure 6.9 can capture the increased

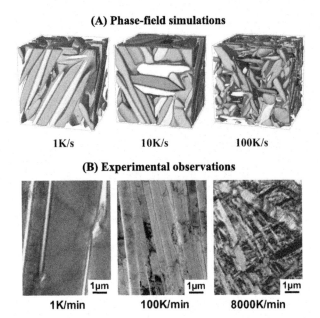

(A) Phase-field simulations

1K/s   10K/s   100K/s

(B) Experimental observations

1µm   1µm   1µm

1K/min   100K/min   8000K/min

FIGURE 6.9   Phase-field simulation of $\beta \to \alpha$ transformations under different cooling rates in β single crystal and the comparison with experimental observations: (A) phase-field simulations, system size 10 µm × 10 µm × 10 µm; (B) experimental observations [40] (Credit: G. Lütjering).

α number density and decreased α size with increasing cooling rates, which qualitatively agrees with the experimental observations [40]. Based on the phase-field simulations, the microstructure evolution during $\beta \to \alpha$ transformations under different cooling rates is governed by the competition between nucleation and growth. Under high cooling rates (100 K/s), there is an enhanced nucleation rate due to the high undercooling, which results in larger nucleation rate; on the other hand, the growth of the α nuclei is largely retarded, due to the decreased growth rates and increased competitions among different α nuclei (since the increased nucleation rate would lead to significant decrease in average spacing of α nuclei) at low temperatures. Under low cooling rates (1 K/s), the system can stay at the high-temperature regime for a sufficiently long time, so that the α nuclei can grow larger, which, on the other hand, retards the further nucleation of α products.

For a specific RVE in the macroscopic thermal model, its thermal history during AM involves multiple cooling/heating cycles with non-constant cooling/heating rates. Based on the thermal history calculations, the cooling/heating rates increase significantly when the RVE is near the heat source, and the overall effect for the multiple thermal cycles is a cooling process, as shown in Figure 6.3B. Therefore, in a RVE of the thermal model, below the β transus, after each cooling/heating cycle, the α products will form and dissolve, while the overall effect is the formation of certain amount of α products without dissolution. The α products left over after a cooling/heating cycle are generally the ones formed at the earlier stages during the previous cooling process, which will also affect the formation of subsequent α products. Therefore, it is critical to understand the formation and growth sequence

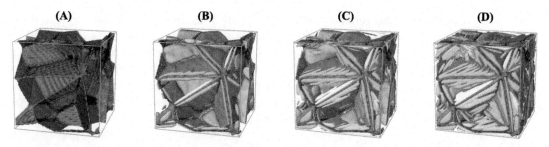

FIGURE 6.10 Sequential formation of α products during slow cooling of a polycrystal: (A) initial grain structure; (B) formation of GB-α; (C) development of α-colonies; (D) basket-weave+colony microstructure. System size: 10 μm × 10 μm × 10 μm.

of α products during cooling, especially in polycrystals. As mentioned by Kelly and Ji et al. [13,90], during early stages of cooling, the driving force for nucleating an α nucleus is limited, resulting in low nucleation rate in grain interiors. In this case, grain boundaries and/or pre-existing dislocations can facilitate the nucleation of α particles through heterogeneous nucleation, resulting in grain boundary α (GB-α) products as the initially observed α products, especially during low cooling rates. The GB-α products are usually of plate shapes parallel to the GB and form protrusions which develop into α-colonies consisting of the same α variants parallel to each other. The formation of α-colonies is either due to the instabilities during the growth of GB-α or the elastic interactions between the α nuclei and pre-existing GB-α plate, as has been discussed in [91]. During the growth of α-colonies, new α nuclei can form inside β grains to form basket-weave-type microstructures. To illustrate this process, the microstructure evolution during slow cooling in a Ti-6Al-4V polycrystal is simulated, as shown in Figure 6.10, which captures the sequential microstructure evolution features. Based on the current simulation results, further simulations will be performed in Ti-6Al-4V polycrystals by applying the predicted thermal history during AM from the macroscopic thermal model and the initial grain structure from the grain growth model. Furthermore, the effect of AM parameters on the microstructure development will be investigated.

6.2.4.3 Future Directions

The sub-grain microstructure evolution varies with the specific material systems. The current sub-grain phase-field model focuses on the $\beta \to \alpha$ transformations, which is the major cause for microstructure evolution during AM of Ti-6Al-4V. Specifically for this system, the phase-field model has been extended for non-isothermal simulations. However, it is still numerically challenging to consider the realistic thermal history of a RVE using the current model due to the rapid thermal cycles (e.g., cooling/heating rates ~5000 K/s in Figure 6.3B). The extension of the current model to these extreme cases would definitely be worthwhile pursuing for the understanding of sub-grain microstructure evolution during AM.

For other materials systems, the phase transformations in sub-grain scale can become more complicated. For example, during AM of superalloy IN718, different types of precipitates can appear at GBs and/or inside grains, including the Laves phase, δ-Ni$_3$Nb, γ'-Ni$_3$Al and γ''-Ni$_3$Nb precipitates, which have different effects on the mechanical properties of the build. Specifically, the intra-granular γ' and γ'' precipitates, either sequentially precipitated or co-precipitated, can remain coherent with the γ-matrix and provide notable precipitate hardening [92]; the GB-δ can partially enhance the mechanical property of the build by impeding grain growth, while the intragranular δ, which has limited coherency with γ-matrix, is detrimental to the overall strength of the build [92]. To simulate the microstructure evolution of these precipitate phases, the sub-grain phase-field model should be further extended to multi-component, multi-phase systems coupled with corresponding temperature-dependent thermodynamic database and diffusion kinetic information.

6.3 SUMMARY AND OUTLOOK

In summary, the proposed multi-scale phase-field modeling framework has been shown to be capable of capturing the major grain-scale and sub-grain scale microstructural features during the AM of Ti-6Al-4V. The effect of temperature distribution and thermal history on microstructural evolution mechanisms have been investigated, which can be further correlated with the macroscopic AM processing parameters. The possibility and capability of constructing such an integrated computational framework are also shown to help further understanding and future predictions of microstructural evolution during the AM processes. The correlation between AM processing parameters and the resulting microstructures during metallic AM can also be investigated using the techniques presented in this work. For individual models, qualitative agreements have been achieved between the simulation results and experimental observations, including grain structures in the longitudinal and wall sections of the build, as well as the habit planes and morphologies of α phases in β grains and near β grain boundaries. More accurate and reliable predictions of the microstructure evolution would rely on the improvement, extension and validation of the current model, especially in the below described directions.

(1) The numerical accuracy, robustness and efficiency should be quantified and further improved. This does not only include the numerical improvement for the three sub-models individually, the macroscopic thermal model, grain-scale grain growth/solidification phase-field model and sub-grain scale phase-field model for solid state phase transformation, so that the models can consider more extreme situations during realistic AM processes; but also include the improvement for the connect-interface among the three models, so that the key information related to the AM process from the larger-scale model can be input into the smaller-scale model without much loss in numerical accuracy.

(2) The quantitative validation of the predicted microstructure morphology during AM should be performed with existing experimental results. This can be accomplished by performing parallel AM experiments with the same material and processing parameters as that in the phase-field model. Model validation can be achieved by characterizing the as-built sample using different techniques in a quantitative manner, such as EBSD for grain texture and variant distribution and SEM/TEM for microstructure morphology analysis. By quantitatively comparing the simulation results and experimental observations, new insight can be obtained for better improving the simulation methods, numerical treatments and model parameters.

(3) The accuracy of phase-field simulations is also related to the accuracy of the input materials parameters, as well as the temperature-dependent thermodynamic and diffusional mobility databases of the material system. These largely rely on more accurate experimental calibrations. For the database development, more experiments on phase boundary identification and inter-diffusion measurement are desired. First-principles calculations on the relative phase stabilities, self-diffusivities and impurity diffusivities can also be an important contribution.

References

[1] Babu SS, Love L, Dehoff R, Peter W, Watkins TR, Pannala S. Additive manufacturing of materials: opportunities and challenges. Mater Res Soc Bull 2015;40:1154–61.

[2] Körner C. Additive manufacturing of metallic components by selective electron beam melting — a review. Int Mater Rev 2016;61:361–77.

[3] Basak A, Das S. Epitaxy and microstructure evolution in metal additive manufacturing. Annu Rev Mater Res 2016;46:125–49.

[4] Gu DD, Meiners W, Wissenbach K, Poprawe R. Laser additive manufacturing of metallic components: materials, processes and mechanisms. Int Mater Rev 2013;57:133–64.

[5] King WE, Anderson AT, Ferencz RM, Hodge NE, Kamath C, Khairallah SA, Rubenchik AM. Laser powder bed fusion additive manufacturing of metals, physics, computational, and materials challenges. Appl Phys Rev 2015;2:041304.

[6] Lewandowski JJ, Seifi M. Metal additive manufacturing: a review of mechanical properties. Annu Rev Mater Res 2016;46:151–86.

[7] Frazier WE. Metal additive manufacturing: a review. J Mater Eng Perform 2014;23:1917–28.

[8] Sames WJ, List FA, Pannala S, Dehoff RR, Babu SS. The metallurgy and processing science of metal additive manufacturing. Int Mater Rev 2016;61:315–60.

[9] Collins PC, Brice DA, Samimi P, Ghamarian I, Fraser HL. Microstructural control of additively manufactured metallic materials. Annu Rev Mater Res 2016;46:63–91.

[10] Markl M, Körner C. Multiscale modeling of powder bed–based additive manufacturing. Annu Rev Mater Res 2016;46:93–123.

[11] Beese AM, Carroll BE. Review of mechanical properties of Ti-6Al-4V made by laser-based additive manufacturing using powder feedstock. JOM 2015;68:724–34.

[12] Yap CY, Chua CK, Dong ZL, Liu ZH, Zhang DQ, Loh LE, Sing SL. Review of selective laser melting: materials and applications. Appl Phys Rev 2015;2:041101.

[13] Kelly SM. Thermal and microstructure modeling of metal deposition processes with application to Ti-6Al-4V. Virginia Polytechnic Institute and State University; 2004.

[14] Chen L-Q. Phase-field models for microstructure evolution. Annu Rev Mater Res 2002;32:113–40.

[15] Xu W, Brandt M, Sun S, Elambasseril J, Liu Q, Latham K, Xia K, Qian M. Additive manufacturing of strong and ductile Ti-6Al-4V by selective laser melting via in situ martensite decomposition. Acta Mater 2015;85:74–84.

[16] Carroll BE, Palmer TA, Beese AM. Anisotropic tensile behavior of Ti-6Al-4V components fabricated with directed energy deposition additive manufacturing. Acta Mater 2015;87:309–20.

[17] Antonysamy AA, Meyer J, Prangnell PB. Effect of build geometry on the β-grain structure and texture in additive manufacture of Ti-6Al-4V by selective electron beam melting. Mater Charact 2013;84:153–68.

[18] Hernández-Nava E, Smith CJ, Derguti F, Tammas-Williams S, Leonard F, Withers PJ, Todd I, Goodall R. The effect of defects on the mechanical response of Ti-6Al-4V cubic lattice structures fabricated by electron beam melting. Acta Mater 2016;108:279–92.

[19] Cunningham R, Narra SP, Ozturk T, Beuth J, Rollett AD. Evaluating the effect of processing parameters on porosity in electron beam melted Ti-6Al-4V via synchrotron X-ray microtomography. JOM 2016;68:765–71.

[20] Qiu CL, Ravi GA, Dance C, Ranson A, Dilworth S, Attallah MM. Fabrication of large Ti-6Al-4V structures by direct laser deposition. J Alloys Compd 2015;629:351–61.

[21] Sterling AJ, Torries B, Shamsaei N, Thompson SM, Seely DW. Fatigue behavior and failure mechanisms of direct laser deposited Ti-6Al-4V. Mater Sci Eng A 2016;655:100–12.

[22] Edwards P, Ramulu M. Fatigue performance evaluation of selective laser melted Ti-6Al-4V. Mater Sci Eng 2014;598:327–37.

[23] Åkerfeldt PAM, Pederson R. Influence of microstructure on mechanical properties of laser metal wire-deposited Ti-6Al-4V. Mater Sci Eng 2016;674:428–37.

[24] Kobryn PA, Semiatin SL. The laser additive manufacture of Ti-6Al-4V. JOM 2001;53:40–2.

[25] Lu SL, Qian M, Tang HP, Yan M, Wang J, St John DH. Massive transformation in Ti-6Al-4V additively manufactured by selective electron beam melting. Acta Mater 2016;104:303–11.

[26] Lin JJ, Lv YH, Liu YX, Xu BS, Sun Z, Li ZG, Wu YX. Microstructural evolution and mechanical properties of Ti-6Al-4V wall deposited by pulsed plasma arc additive manufacturing. Mater Des 2016;102:30–40.

[27] Wang F, Williams S, Colegrove P, Antonysamy AA. Microstructure and mechanical properties of wire and arc additive manufactured Ti-6Al-4V. Metall Mater Trans A 2012;44:968–77.

[28] Denlinger ER, Heigel JC, Michaleris P. Residual stress and distortion modeling of electron beam direct manufacturing Ti-6Al-4V. Proc Inst Mech Eng, B J Eng Manuf 2014;229:1803–13.

[29] Keist JS, Palmer TA. Role of geometry on properties of additively manufactured Ti-6Al-4V structures fabricated using laser based directed energy deposition. Mater Des 2016;106:482–94.

[30] Thijs L, Verhaeghe F, Craeghs T, Van Humbeeck J, Kruth J. A study of the micro structural evolution during selective laser melting of Ti-6Al-4V. Acta Mater 2010;58:3303–12.

[31] Sridharan N, Chaudhary A, Nandwana P, Babu SS. Texture evolution during laser direct metal deposition of Ti-6Al-4V. JOM 2016;68:772–7.

[32] Gong H, Rafi K, Gu H, Janaki Ram GD, Starr T, Stucker B. Influence of defects on mechanical properties of Ti-6Al–4V components produced by selective laser melting and electron beam melting. Mater Des 2015;86:545–54.

[33] Vrancken B, Thijs L, Kruth J, Van Humbeeck J. Heat treatment of Ti-6Al-4V produced by selective laser melting microstructure and mechanical properties. J Alloys Compd 2012;541:177–85.

[34] Yadroitsev I, Krakhmalev P, Yadroitsava I. Selective laser melting of Ti-6Al-4V alloy for biomedical applications temperature monitoring and microstructural evolution. J Alloys Compd 2014;583:404–9.

[35] Åkerfeldt P, Antti M, Pederson R. Influence of microstructure on mechanical properties of laser metal wire-deposited Ti-6Al-4V.pdf. Mater Sci Eng A 2016;674:428–37.

[36] de Formanoir C, Michotte S, Rigo O, Germain L, Godet S. Electron beam melted Ti-6Al-4V: microstructure, texture and mechanical behavior of the as-built and heat-treated material. Mater Sci Eng 2016;652:105–19.

[37] Antonysamy AA. Mechanical property evolution during additive manufacturing of Ti-6Al-4V alloy for aerospace applications. University of Manchester; 2012.

[38] Ahmed T, Rack HJ. Phase transformations during cooling in $\alpha + \beta$ titanium alloys. Mater Sci Eng 1998;243:206–11.

[39] Lütjering G. Influence of processing on microstructure and mechanical properties of $(\alpha + \beta)$ titanium alloys. Mater Sci Eng 1998;243:32–45.

[40] Lütjering G, Williams JC. Titanium. 2nd edn. Berlin, Heidelberg: Springer; 2007.

[41] Traini T, Mangano C, Sammons RL, Mangano F, Macchi A, Piattelli A. Direct laser metal sintering as a new approach to fabrication of an isoelastic functionally graded material for manufacture of porous titanium dental implants. Dent Mater 2008;24:1525–33.

[42] Manvatkar V, De A, DebRoy T. Heat transfer and material flow during laser assisted multi-layer additive manufacturing. J Appl Phys 2014;116:124905.

[43] Raghavan A, Wei HL, Palmer TA, DebRoy T. Heat transfer and fluid flow in additive manufacturing. J Laser Appl 2013;25:052006.

[44] Gouge MF, Heigel JC, Michaleris P, Palmer TA. Modeling forced convection in the thermal simulation of laser cladding processes. Int J Adv Manuf Technol 2015;79:307–20.

[45] Michaleris P. Modeling metal deposition in heat transfer analyses of additive manufacturing processes. Finite Elem Anal Des 2014;86:51–60.

[46] Yan W, Ge W, Smith J, Lin S, Kafka OL, Lin F, Liu WK. Multi-scale modeling of electron beam melting of functionally graded materials. Acta Mater 2016;115:403–12.

[47] Smith J, Xiong W, Cao J, Liu WK. Thermodynamically consistent microstructure prediction of additively manufactured materials. Comput Mech 2016;57:359–70.

[48] Denlinger ER, Irwin J, Michaleris P. Thermomechanical modeling of additive manufacturing large parts. J Manuf Sci Eng 2014;136:061007.

[49] Nie P, Ojo OA, Li Z. Numerical modeling of microstructure evolution during laser additive manufacturing of a nickel-based superalloy. Acta Mater 2014;77:85–95.

[50] Zhou X, Zhang H, Wang G, Bai X, Fu Y, Zhao J. Simulation of microstructure evolution during hybrid deposition and micro-rolling process. J Mater Sci 2016;51:6735–49.

[51] Wei HL, Elmer JW, DebRoy T. Origin of grain orientation during solidification of an aluminum alloy. Acta Mater 2016;115:123–31.

[52] Van Ginneken AJJ, Burgers WG. The habit plane of the zirconium transformation. Acta Crystallogr 1952;5:548–9.

[53] Emmerich H. Advances of and by phase-field modelling in condensed-matter physics. Adv Phys 2008;57:1–87.

[54] Steinbach I. Phase-field model for microstructure evolution at the mesoscopic scale. Annu Rev Mater Res 2013;43:89–107.

[55] Furrer DU. Application of phase-field modeling to industrial materials and manufacturing processes. Curr Opin Solid State Mater Sci 2011;15:134–40.

[56] Wang Y, Li J. Phase field modeling of defects and deformation. Acta Mater 2010;58:1212–35.

[57] Singer-Loginova I, Singer HM. The phase field technique for modeling multiphase materials. Rep Prog Phys 2008;71:106501.

[58] Nestler B, Choudhury A. Phase-field modeling of multi-component systems. Curr Opin Solid State Mater Sci 2011;15:93–105.

[59] Heo TW, Chen L-Q. Phase-field modeling of nucleation in solid-state phase transformations. JOM 2014;66:1520–8.

[60] Steinbach I. Phase-field models in materials science. Model Simul Mater Sci Eng 2009;17:073001.

[61] Boettinger WJ, Warren JA, Beckermann C, Karma A. Phase-field simulation of solidification. Annu Rev Mater Res 2002;32:163–94.

[62] Gong XB, Chou K. Phase-field modeling of microstructure evolution in electron beam additive manufacturing. JOM 2015;67:1176–82.

[63] Lim H, Abdeljawad F, Owen SJ, Hanks BW, Foulk JW, Battaile CC. Incorporating physically-based microstructures in materials modeling: bridging phase field and crystal plasticity frameworks. Model Simul Mater Sci Eng 2016;24:045016.

[64] Abaqus. V6.13 edn. Providence, RI: USA; 2013.

[65] Krill III CE, Chen L-Q. Computer simulation of 3-D grain growth using a phase-field model. Acta Mater 2002;50:3057–73.

[66] Chen L, Chen J, Lebensohn RA, Ji YZ, Heo TW, Bhattacharyya S, Chang K, Mathaudhu S, Liu ZK, Chen LQ. An integrated fast Fourier transform-based phase-field and crystal plasticity approach to model recrystallization of three dimensional polycrystals. Comput Methods Appl Mech Eng 2015;285:829–48.

[67] Dinda GP, Dasgupta AK, Mazumder J. Texture control during laser deposition of nickel-based superalloy. Scr Mater 2012;67:503–6.

[68] Burgers WG. On the process of transition of the cubic-body-centered modification into the hexagonal-close-packed modification of zirconium. Physica 1934;1:561–86.

[69] Wang D, Shi R, Zheng Y, Banerjee R, Fraser HL, Wang Y. Integrated computational materials engineering (ICME) approach to design of novel microstructures for Ti-alloys. JOM 2014;66:1287–98.

[70] Shi R, Ma N, Wang Y. Predicting equilibrium shape of precipitates as function of coherency state. Acta Mater 2012;60:4172–84.

[71] Boyne A, Wang D, Shi R, Zheng Y, Behera A, Nag S, Tiley JS, Fraser HL, Banerjee R, Wang Y. Pseudospinodal mechanism for fine α/β microstructures in β-Ti alloys. Acta Mater 2014;64:188–97.

[72] Shi R, Wang Y. Variant selection during α precipitation in Ti-6Al-4V under the influence of local stress – a simulation study. Acta Mater 2013;61:6006–24.

[73] Shi R, Dixit V, Fraser HL, Wang Y. Variant selection of grain boundary α by special prior β grain boundaries in titanium alloys. Acta Mater 2014;75:156–66.

[74] Zhang F. Thermodynamic database of Ti-Al-V ternary system. Madison, WI, USA; 2013. LLC, C., Ed.

[75] Li D, Zhu L, Shao S, Jiang Y. First-principles based calculation of the macroscopic α/β interface in titanium. J Appl Phys 2016;119:225302.

[76] Kobayashi H, Ode M, Kim SG, Kim WT, Suzuki T. Phase-field model for solidification of ternary alloys coupled with thermodynamic database. Scr Mater 2003;48:689–94.

[77] Elmer JW, Palmer TA, Babu SS, Specht ED. In situ observations of lattice expansion and transformation rates of α and β phases in Ti–6Al–4V. Mater Sci Eng A 2005;391:104–13.

[78] Khachaturyan AG. Theory of structural transformations in solids. Mineola, New York: Dover Publications, Inc.; 2008.

[79] Huang L, Cui Y, Chang H, Zhong H, Li J, Zhou L. Assessment of atomic mobilities for bcc phase of Ti-Al-V system. J Phase Equilibria Diffus 2010;31:135–43.

[80] Xu WW, Shang SL, Zhou BC, Wang Y, Chen LJ, Wang CP, Liu XJ, Liu ZK. A first-principles study of diffusion coefficients of alloying elements in dilute α-Ti alloys. Phys Chem Chem Phys 2016;18:16870–81.

[81] Heo TW, Bhattacharyya S, Chen L-Q. A phase-field model for elastically anisotropic polycrystalline binary solid solutions. Philos Mag 2013;93:1468–89.

[82] Newkirk JB, Geisler AH. Crystallographic aspects of the β to α transformation in titanium. Acta Metall 1953;1:371–4.

[83] McHargue CJ. Crystallographic aspects of the β to α transformation in titanium. Acta Crystallogr 1953;6:529–30.

[84] Williams AJ, Cahn RW, Barrett CS. The crystallography of the $\beta - \alpha$ transformation in titanium. Acta Metall 1954;2:117–28.

[85] Furuhara T, Nakamori H, Maki T. Crystallography of α phase precipitated on dislocations and deformation twin boundaries in a β titanium alloy. Mater Trans, JIM 1992;33:585–95.

[86] Mills MJ, Hou DH, Suri S, Viswanathan GB. Orientation relationship and structure of alpha/beta interface in conventional titanium alloys. In: Pond RC, Clark WAT, King AH, editors. Boundaries and interfaces in materials: the David A. Smith symposium. The Minerals, Metals & Materials Society; 1998. p. 295.

[87] Simmons JP, Wen Y, Shen C, Wang YZ. Microstructural development involving nucleation and growth phenomena simulated with the phase field method. Mater Sci Eng A 2004;365:136–43.

[88] Simmons JP, Shen C, Wang Y. Phase field modeling of simultaneous nucleation and growth by explicitly incorporating nucleation events. Scr Mater 2000;43:935–42.

[89] Kozeschnik E, Holzer I, Sonderegger B. On the potential for improving equilibrium thermodynamic databases with kinetic simulations. J Phase Equilibria Diffus 2007;28:64–71.

[90] Ji Y, Heo TW, Zhang F, Chen L-Q. Theoretical assessment on the phase transformation kinetic pathways of multi-component Ti alloys: application to Ti-6Al-4V. J Phase Equilibria Diffus 2015;37:53–64.

[91] Radhakrishnan B, Gorti S, Babu SS. Phase field simulations of autocatalytic formation of alpha lamellar colonies in Ti-6Al-4V. Metall Mater Trans A 2016.

[92] Sundararaman M, Mukhopadhyay P, Banerjee S. Some aspects of the precipitation of metastable intermetallic phases in INCONEL 718. Metall Trans A 1992;23A:2015–28.

Modeling Microstructure of AM Processes Using the FE Method*

Jeff Irwin

Product Development Group, Autodesk Inc., State College, USA

7.1 INTRODUCTION

Out of all the titanium alloys in use, Ti-6Al-4V is the most widely produced [1]. This alloy is commonly used in aerospace applications because of its light weight, high strength, corrosion resistance, and because it maintains these characteristics at high temperatures [2]. At room temperature under equilibrium, Ti-6Al-4V is primarily composed of a hexagonal close packed α phase, and to a lesser extent, a body centered cubic β phase. At higher temperatures, approaching the so-called β transus temperature, the α phase transforms to the β phase. Upon cooling back to room temperature, the β phase transforms back to the primary α phase for low cooling rates, or to a martensitic α' phase for high cooling rates. The morphology of these phases varies significantly, depending on both thermal and mechanical processing, leading to corresponding changes in mechanical properties [3–6].

Early theoretical work on evolution of microstructures as a function of time and temperature was done by Avrami [7–9]. He proposed a two-stage transformation, where germs of the new phase are first nucleated, then proceed to grow in volume. According to his theory, the rates at which these two stages progress are governed by temperature. Johnson, Mehl, and Kolmogorov [10,11] performed similar work contemporaneous to Avrami. More recently, Malinov et al. [12] investigated the phenomena of α growth in Ti-6Al-4V upon cooling from the β phase field. They measured the kinetic parameters needed to realize the Johnson–Mehl–Avrami (JMA) model for this alloy system, but only under isothermal conditions.

Directed Energy Deposition (DED) is an additive manufacturing (AM) process that operates by delivering a wire or powdered material into a molten pool formed by a concentrated heat source, such as a laser or electron beam. As the heat source and powder delivery nozzle move relative to the workpiece in the build plane, a layer of any shape can be built. At

---

* This chapter is based upon the original work: Jeff Irwin, Edward W. Reutzel, Pan Michaleris, Jay Keist, and Abdalla R. Nassar, "Predicting Microstructure from Thermal History during Additive Manufacturing for Ti-6Al-4V." J Manuf Sci Eng 138(11) (2016): 111007-1 to 111007-11.

the completion of a layer, the nozzle increments upwards and a new layer is begun. In this way, a part with a complex 3D geometry is built layer by layer. AM is being used as a means of repairing expensive parts [13], for fabrication of custom bio-medical components [14], for adding features to components, and for fabricating preforms. Unfortunately, AM is prone to failure due to residual stress or distortion, and parts are currently designed by trial and error [15–20].

Several researchers have implemented models of the AM process [21–24]. Of particular note is Kelly, who [21] performed extensive work, both experimental [25] and computational [26], on the microstructure of Ti-6Al-4V produced by an AM process. He used Christian's principle of additivity [27] to apply the isothermal JMA model to an anisothermal process. However, Kelly's microstructure model of the LENS® system, a powder-fed DED process [28], was based on a simple finite-difference thermal model. His 2D model used a structured mesh, limiting the model to simple geometries. Unlike finite-difference models, finite element models are capable of using elements of any shape and connectivity. The flexibility of the available shapes is twofold. First, more complex geometries can be meshed. Second, the mesh can be coarsened more effectively, reducing the number of degrees of freedom and the computational expense.

Fan et al. and Crespo coupled microstructure models with more advanced 3D thermo-mechanical Finite Element Method (FEM) models of a laser forming process [29] and a 75 layer laser powder deposition process [30]. Fan [29] showed that accounting for the effects of phase transformations on mechanical properties resulted in more accurate prediction of bending angles. However, the laser forming process involves only a single laser pass and is much less complex to model compared to a multi-layer deposition. Crespo [30] used X-ray diffraction to validate the β phase fraction predicted by his model and depth sensing indentation to validate the predicted Young's modulus and hardness. However, the comparison was only made at 3 different measurement points and the error was not quantified. Charles et al. performed a thermo-microstructural simulation of Tungsten inert gas metal wire deposition for single bead walls, ranging in height from 8 to 30 layers [31,32]. Similar to Kelly, she used a JMA model for α growth. Unlike Kelly, Charles did not consider the kinetics of α dissolution in her original work, but included martensite α phase fraction and α lath width in her analysis. Further, she successfully validated the thermal model with thermocouple measurements. In a later work, Charles does account for the full α dissolution kinetics [33].

The repeated thermal cycling inherent in layer-wise AM processing makes the microstructural evolution sensitive to kinetic parameters and other material properties. This cycling is caused by the deposition of many subsequent laser passes and layers. Although Kelly successfully validated his model with in-situ X-ray diffraction measurements during several thermal cycles induced by resistance heating, he was unable to obtain more than a qualitative agreement with experimental results for an 18 layer AM part [21]. An accurate microstructural model is needed in order to predict and assess the quality of DED and other AM parts from simulated or measured temperature history. Such a model may be used to develop so-called thermal metrics, i.e. a method to measure characteristics of the thermal history that can be reduced to a small set of scalar values that correlate to features in the microstructure. Thermal metrics are "quantities that represent characteristics of the build related to the temperature field during processing" [28] such as melt pool surface area and α lath width. Many researchers have noted the strong dependence of the mechanical properties of Ti-6Al-4V on its microstructure [3–6,34]. In turn, these mechanical properties must be known in order to develop accurate mechanical models of AM processes and to certify that finished AM parts satisfy their design requirements. While thermo-microstructural models of AM processes can be performed without mechanical properties, models of residual stress and distortion require elastic modulus, yield strength, etc. Given a 10 GB database of temperature history (typical for IR imaging [28] or FEA models), it is difficult to determine whether a part will attain its desired mechanical properties, hence the need for costly trial and error design. Thermal metrics have the capability of reducing this amount of data by 4 orders of magnitude [28] and are easier to interpret for the purpose of part certification. These metrics will allow simulations to replace costly trial and error iterations.

Inverse simulation is a common technique to estimate unknown material properties [35] or to improve the performance of a control system [36]. It is alternatively referred to as inverse analysis, model calibration, or model optimization [37]. These methods work by varying any unknown properties or parameters until a better agreement with experimental results is achieved.

This work presents a novel implementation of Kelly's microstructure model and Charles's lath width model, and is experimentally validated with two complex large builds. An inverse simulation is performed to improve the accuracy of the model, and simulation-based thermal histories are used as input. The combined Kelly–Charles model is reviewed in §7.2. A microstructural characterization of two 286 layer, complex DED parts with different interlayer dwell times is carried out as described in §7.3, and the results of the new model are compared to the Kelly–Charles model and experimental results in §7.4.

7.2 MICROSTRUCTURAL MODEL

7.2.1 Phase Fractions and Morphology

The phase fraction and morphological model follows Kelly's work [21]. Given the thermal history for a material point in Ti-6Al-4V, this model calculates the α phase fraction f_α and the β phase fraction f_β. The α phase fraction is further divided into the colony-α fraction $f_{C-\alpha}$ and the basketweave-α fraction $f_{BW-\alpha}$. These phase fractions obey the following relations:

$$f_\alpha + f_\beta = 1 \tag{7.1}$$

$$f_{C-\alpha} + f_{BW-\alpha} = f_\alpha \tag{7.2}$$

The model loops through a series of discrete times $t_0, t_1, \ldots, t_i$ with corresponding temperatures $T_0, T_1, \ldots, T_i$ calculated by Netfabb Simulation. At each time t_i, the equilibrium α phase fraction $f_{eq,\alpha}(T_i)$ is calculated by a table lookup [21] based on the temperature T_i, as shown in Figure 7.1. If the current fraction f_α predicted by the model is greater than the equilibrium fraction $f_{eq,\alpha}$, then some of the α phase dissolves, as described in §7.2.1.1. Alternatively, if f_α is less than $f_{eq,\alpha}$, then additional α phase will nucleate and/or grow, as described in §7.2.1.2. Martensitic α' is not included in Equation (7.2), as discussed in §7.2.4. The initial conditions for the simulation are set as 0.09 β fraction, 0.91 colony-α fraction, and 0 basketweave-α. The

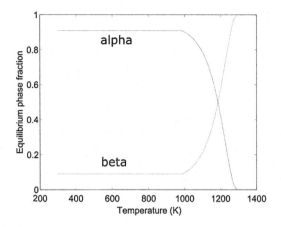

FIGURE 7.1 Equilibrium α and β phase fractions from Ref. [21].

β initial condition corresponds to its room temperature equilibrium value, while the colony and basketweave α initial conditions are somewhat arbitrary. The bulk of the substrate is not heated enough to change from these initial conditions in the model, while the initial microstructure in the deposition is erased as it crosses the β transus temperature. Hence, the colony and basketweave α initial conditions do not have any effect on the final results in the deposition.

7.2.1.1 α Dissolution

Dissolution of the α phase is modeled as one-dimensional plate growth of the β phase, where the reaction proceeds at a parabolic rate [27]. First, the parabolic reaction rate r [s$^{-1/2}$] is calculated from the current temperature T_i as

$$r = a \left(\frac{T_i}{T_{\text{ref}}} \right)^b \tag{7.3}$$

where the coefficient $a = 2.21 \times 10^{-31}$ s$^{-1/2}$ and the exponent $b = 9.89$ (dimensionless) were calculated by Kelly using ThermoCalc® and DiCTra® simulations [21]. The reference temperature, included for dimensional consistency, is taken as $T_{\text{ref}} = 1$ K. This expression for r is based on a simplified one-dimensional steady-state case of diffusion-controlled isothermal β growth models. The equivalent time for dissolution t_D^* is the amount of time necessary at the current temperature, starting from 100% α, to reach the amount of α present at the previous time step $f_{i-1,\alpha}$, and is calculated as

$$t_D^* = \left(\frac{f_{i-1,\beta}}{f_{\text{eq},\beta} r} \right)^2 \tag{7.4}$$

where $f_{i-1,\beta}$ is the β fraction from the previous time step, and the equilibrium β fraction is given by $f_{\text{eq},\beta} = 1 - f_{\text{eq},\alpha}$. Finally, the extent of reaction ζ_β and the updated β phase fraction

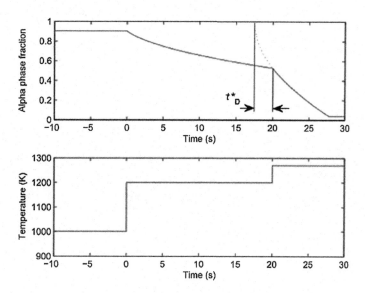

FIGURE 7.2 An illustration of the concept of equivalent time.

$f_{i,\beta}$ are calculated as

$$\zeta_\beta = r\sqrt{\Delta t + t_D^*} \tag{7.5}$$

$$f_{i,\beta} = f_{eq,\beta}\zeta_\beta \tag{7.6}$$

where the time step is $\Delta t = t_i - t_{i-1}$, the updated α phase fraction is calculated in accordance with Equation (7.1), and ζ_β is bounded in the range [0, 1]. If the extent of reaction ζ_β is 0, then the α phase does not dissolve and no β is present. If ζ_β is 1, then the reaction proceeds fully to equilibrium. When ζ_β is between these extreme values, the reaction takes a step towards the equilibrium phase fraction $f_{eq,\beta}$, but does not fully attain $f_{eq,\beta}$ during the time step. To further explain equivalent time, a simple example of α dissolution is shown in Figure 7.2. In this example, the material is in equilibrium at 1000 K until time $t = 0$ s. Then, a step input abruptly raises the temperature to 1200 K. The temperature remains constant until $t = 20$ s when another step input raises the temperature to 1270 K. The corresponding α phase fraction, shown as a solid red curve, dissolves proportionally to the square root of time. At $t = 20$ s, the rate of dissolution is discontinuous due to the temperature dependence of the parabolic reaction rate r. A dotted curve is extended backwards, as if the reaction had started at 1270 K with 100% α slightly before 20 s, to represent the equivalent time for dissolution t_D^*. Finally, the α fraction reaches a new equilibrium at 28 s and stops dissolving.

Whether the dissolved α is taken from the colony or the basketweave morphology depends on the intragranular nucleation temperature $T_{IG} = 1100$ K [21]. If $T_i < T_{IG}$, then only the basketweave morphology dissolves (as long as it does not become negative). If $T_i \geq T_{IG}$, then both basketweave and colony dissolve in the same proportion.

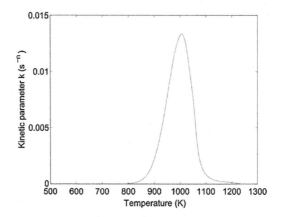

FIGURE 7.3   Kinetic parameter for α growth from Ref. [21].

7.2.1.2 α Growth

The nucleation and growth of the α phase is modeled by a Johnson–Mehl–Avrami model [7–10]. First, the kinetic parameter $k\,(T_i)$ is calculated by a table lookup based on the temperature T_i as shown in Figure 7.3. Kelly calculated this tabular data with time-temperature-transformation diagrams in JMatPro [21]. The equivalent time for growth t_G^* is the amount of time necessary at the current temperature, starting from 0% α, to reach the amount of α present at the previous time step $f_{i-1,\alpha}$, and is calculated as

$$t_G^* = \left[-\frac{1}{k} \log \left(1 - \frac{f_{i-1,\alpha}}{f_{eq,\alpha}} \right) \right]^{1/n} \tag{7.7}$$

where $f_{i-1,\alpha}$ is the α fraction from the previous time step, $n = 5/2$ is the Avrami exponent, and $\log{(\cdot)}$ is the natural logarithm. Finally, the extent of reaction ζ_α and the updated α phase fraction $f_{i,\alpha}$ are calculated as

$$\zeta_\alpha = 1 - \exp\left[-k \left(\Delta t + t_G^* \right)^n \right] \tag{7.8}$$

$$f_{i,\alpha} = f_{eq,\alpha} \zeta_\alpha \tag{7.9}$$

The updated β phase fraction is calculated in accordance with Equation (7.1) and ζ_α is bounded in the range $[0, 1]$. Similar to the morphology rules for α dissolution, if the current temperature is greater than the intragranular nucleation temperature T_{IG}, then the newly grown α becomes part of the colony morphology. Otherwise, it becomes basketweave.

7.2.2 α Lath Width

The α lath width model follows Charles's work [32]. First, the equilibrium value of the α lath width w_{eq} is calculated using an Arrhenius equation

$$w_{eq} = k_w \exp\left(-T_{act}/T_i\right) \tag{7.10}$$

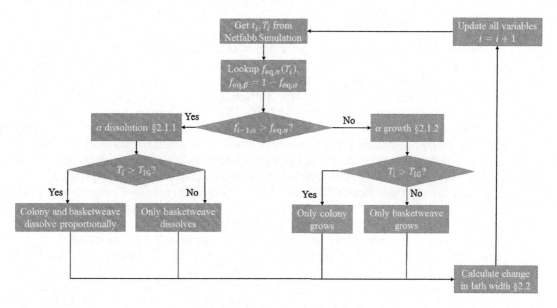

FIGURE 7.4 A flowchart for the thermo-microstructural model.

where the prefactor $k_w = 18\,433$ μm and the activation temperature $T_{act} = 10\,044$ K are taken from Charles's thesis [32], for the case of single bead walls. Then the current α lath width w_i is updated as

$$w_i = \frac{1}{f_{i,\alpha}} \left(w_{i-1} f_{i-1,\alpha} + w_{eq} \Delta f_\alpha \right) \tag{7.11}$$

where w_{i-1} is the lath width from the previous time step and $\Delta f_\alpha = f_{i,\alpha} - f_{i-1,\alpha}$ is the change in α fraction, as calculated by Kelly's model. At the first time step, an initial condition of $w_0 = 1$ μm is used. Similar to the colony and basketweave initial conditions, the lath initial condition has no impact on the final results.

7.2.3 Summary of Model

The logical flow of the thermo-microstructural model is illustrated in Figure 7.4. The process begins at the top, flows to the bottom-right, then iterates in a counter-clockwise manner. At the conclusion of a time step, all variables are updated, control is returned to the thermal analysis in Netfabb Simulation, and the model begins a new time step. The microstructural calculations can be done either concurrently with the thermal analysis, or they can be done as a post-processing step. Alternatively, the thermal simulation can be replaced with experimental measurements (e.g. thermocouples or thermal imaging) by changing only the top-center block of the flowchart.

7.2.4 Model Optimization

The transformation from β to martensitic α' is accounted for by lumping its effect into the transformation $\beta \rightarrow \alpha + \beta$. This approximation is necessary because the martensitic transformation is poorly understood. Particularly, there is a wide discrepancy on the cooling rate necessary for the transformation. Some authors experimentally found that martensite only forms at cooling rates greater than 410 K/s [38], while others found martensite at cooling rates as low as 5.1 K/s [39]. A possible reason for this discrepancy in the literature is that, like basketweave and colony fractions, there is no standard method of quantifying the amount of martensite in a sample. Existing methods for phase fraction characterization, e.g. in Ref. [4], rely on choosing a *threshold value* to convert a grayscale micrograph to a binary image. Choosing an improper threshold value can yield inacurate results. Complicating matters, the microstructural appearance of martensite is visually similar to that of the basketweave morphology. Additionally, α and α' both have HCP structures with comparable lattice constants [40], meaning that X-ray diffraction measurements are unable to distinguish the two phases. This limitation casts doubt on the martensite transformation model of Ref. [30], which uses the X-ray diffraction measurements from Elmer et al. [40] to calculate kinetic parameters. Finally, the martensitic phase is not of interest here, as the only experimentally validated result is α lath width. Lath width depends on the total α content, not the individual basketweave, colony, and martensite phase fractions. While the rate of the martensitic transformation has some impact on the overall α fraction, it is expected to be less important than the primary α/β kinetics.

Lumping the martensite transformation into the primary α transformation is achieved by optimizing the material properties that govern α growth and dissolution. The microstructure model takes several material properties as input, which are either taken from the literature or estimated from other models. These properties govern how quickly the α phase grows (n) and dissolves (a and b) and how the equilibrium lath width varies with temperature (k_w and T_{act}). To obtain an estimate of these properties that takes martensite into account, an inverse simulation is performed using the adaptive Nelder–Mead simplex algorithm (ANMS) [41,42]. ANMS is a derivative-free optimization algorithm that seeks to minimize the error between the modeled and measured lath widths by iteratively varying the material properties. Due to the tight coupling and discrete nature of Equations (7.1)–(7.11), it is difficult to determine the derivatives of lath width with respect to each of the material properties. Because ANMS does not require these derivatives, it is preferred over more common optimization algorithms for this application.

7.3 EXPERIMENTAL IMPLEMENTATION

7.3.1 Deposition Process

Two different Ti-6Al-4V L-shaped parts were built on the Optomec® LENS® MR-7 system: one with continuously deposited material and another with a 4 s dwell time between layers, during which the laser is turned off. In addition to the dwell time difference, the effects of processing parameters are investigated by depositing one leg of each L with a single laser

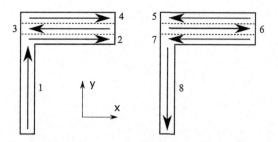

FIGURE 7.5 The order of the eight deposition hatches for odd layers (left) and even layers (right).

FIGURE 7.6 Locations of the cross-sections (dashed lines) for α lath width measurement, superimposed on the 4 s dwell build.

pass at 450 W and the other leg with three laser passes at 350 W. The difference in powers is experimentally determined to achieve a consistent height for each leg of the L. Within the 3-bead leg, the hatch spacing between adjacent laser scans is 0.81 mm. The hatching pattern is illustrated in Figure 7.5. More details regarding the deposition process can be found in Ref. [28].

7.3.2 Measurement of α Lath Width

The cross-sections used for microstructural analysis are from cut ends of the wall structures, as shown in Figure 7.6. The cross-sections are mounted and polished with SiC grinding paper up to 2000 grit size. Final polishing is conducted with 1 μm diamond followed by 0.05 μm colloidal silica. To reveal the microstructure for optical analysis, the samples are etched with Kroll's reagent (3% HNO_3 and 2% HF within water). Optical microscopy is conducted on a Nikon Epiphot microscope. The calibration of the microscope is verified using a stage micrometer.

The α lath thickness measurement is conducted in ImageJ, following the procedures outlined by Tiley et al. [3,4]. First, a threshold image is obtained from the optical micrographs to separate the α and β phases and a grid of parallel lines is overlaid on the image. The parallel lines are then segmented by their intersection with the β phase and the process is repeated by

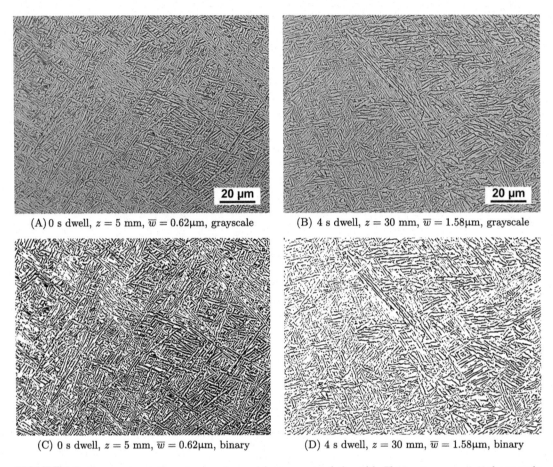

(A) 0 s dwell, $z = 5$ mm, $\overline{w} = 0.62\mu$m, grayscale   (B) 4 s dwell, $z = 30$ mm, $\overline{w} = 1.58\mu$m, grayscale

(C) 0 s dwell, $z = 5$ mm, $\overline{w} = 0.62\mu$m, binary   (D) 4 s dwell, $z = 30$ mm, $\overline{w} = 1.58\mu$m, binary

FIGURE 7.7 Two representative micrographs used for measuring lath width. The top two are original grayscale images, while the lower two are black and white binary images after the thresholding process.

rotating the grid of lines in relation to the threshold image. Finally, the average α lath width in the image $\overline{w}$ is obtained by inverting the line segment lengths λ, calculating the mean inverse value, and using Tiley's relation:

$$\overline{w} = \frac{2}{3\,(1/\lambda)_{\text{mean}}} \tag{7.12}$$

7.4 RESULTS AND DISCUSSION

Several micrographs are shown in Figure 7.7, for a representative thin lath width and thick lath width. At each z height, three images are captured. These three measurements are used to calculate the average and standard deviations of lath width at each height.

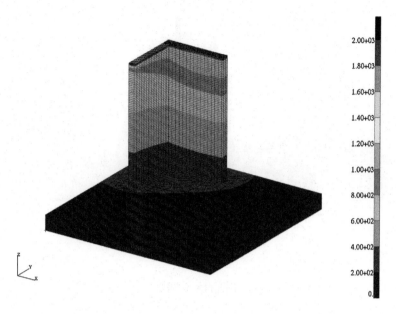

2.00+03
1.80+03
1.60+03
1.40+03
1.20+03
1.00+03
8.00+02
6.00+02
4.00+02
2.00+02
0.

FIGURE 7.8 Temperature profile (°C) at the end of the final laser pass for 0 s dwell.

TABLE 7.1 Material properties suggested by Kelly and Charles, and the optimized values which show better agreement with experimentally measured lath widths.

	Kelly–Charles values	Optimized values
Parabolic coefficient a $(s^{-1/2})$	2.21×10^{-31}	3.07×10^{-31}
Parabolic exponent b (dimensionless)	9.89	9.67
Avrami exponent n (dimensionless)	2.5	2.79
Arrhenius prefactor k_w (µm)	18 400	1.42
Activation temperature T_{act} (K)	10 000	294
12 point calibration error	116%	24.1%
12 point validation error	95.2%	46.4%
Overall 24 point error	106%	37.0%

After performing a thermal analysis in Netfabb Simulation (see Figure 7.8), temperature results are input into the Kelly–Charles model (future work will use measured temperature histories). Using the material properties suggested by Kelly and Charles, the results show poor agreement with experimental measurements of α lath width (106% error). The values of the material properties, before and after optimization, are shown in Table 7.1. Note that the properties that govern α growth and dissolution (a, b, and n) do not change significantly from the published values. On the other hand, k_w and T_{act}, which govern equilibrium α lath widths, are both reduced by several orders of magnitude. The optimized values are physically justified because they are much closer to the length scales and temperatures relevant to the microstructural transformation, respectively µm and hundreds of K.

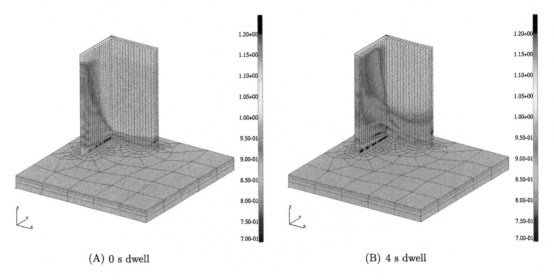

(A) 0 s dwell   (B) 4 s dwell

FIGURE 7.9 α lath width (μm) after the part has cooled.

Lath widths are calculated at all points in Figure 7.9 and compared to experiments in Figure 7.10. Plots of phase fractions, lath width, and temperature are shown in Figure 7.11 as functions of time for a single point in the middle of the 3-bead leg. Error bars on the experimental results are $\pm$ one standard deviation of the three measurements taken at each location. Out of a total 24 experimental data points at different locations and builds, 12 of the points were used for the inverse simulation (Figures 7.10B and 7.10D), while the other 12 were kept hidden from ANMS for validation of the model (Figures 7.10A and 7.10C). Errors for both sets of data are shown in Table 7.1, and are calculated as

$$\varepsilon_{RMS\%} = 100 \times \sqrt{\sum_{\text{data points}} \left(\frac{w_{\text{model}} - w_{\text{experiment}}}{w_{\text{experiment}}} \right)^2} \qquad (7.13)$$

It can be seen that the optimized model gives significantly better results for both sets of data.

Besides those listed in Table 7.1, there are several other material properties. The intragranular nucleation temperature T_{IG} governs how much of the α phase belongs to the colony morphology and how much belongs to the basketweave morphology. However, there is no reliable quantitative method of measuring colony and basketweave content, thus T_{IG} cannot be optimized. Further, the kinetic parameter k and the equilibrium phase fraction $f_{eq,\alpha}$ are temperature dependent. Because of their temperature dependence, optimizing k and $f_{eq,\alpha}$ would require a much larger number of experimental data points.

Although the optimization provides a threefold reduction in error, it may still be improved. A possible source of error is that the temperatures calculated by the thermal model may not be perfect, despite that the thermal model has been validated [43,44], even for a similar build on the LENS® system [45]. For this reason, it might be preferable to use measured temperatures

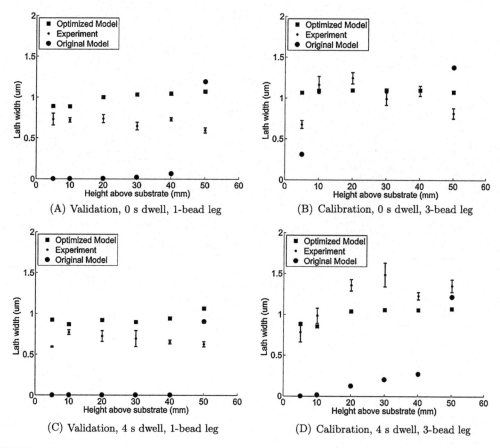

FIGURE 7.10 Experimental lath width measurements compared to the results of two different microstructure models. In Figure 7.10B, the results of the original model are off the scale, with lath widths as large as 3.2 μm.

as input for the microstructure model, rather than coupling one model to another. However, measurements have their own limitations. Besides experimental errors, thermocouples can only measure a small number of points, while infrared imaging can only observe points on the surface. A better long-term strategy would be to improve the accuracy of thermal models through ongoing validation efforts.

Further, it is possible that ANMS converged to a local, rather than a global, minimum. More exhaustive optimization could reduce the error by finding better microstructure material properties. Another source of error is that the martensitic kinetics are lumped into the primary α transformation, rather than being explicitly modeled as a separate phase. This source of error may not be problematic for Kelly and Charles, who operate in different processing regimes where the formation and decomposition of α' are not important. Finally, variations in the exact chemical composition of the alloy make it difficult to compare the original Kelly–Charles model to the optimized model. The formation of microstructure features

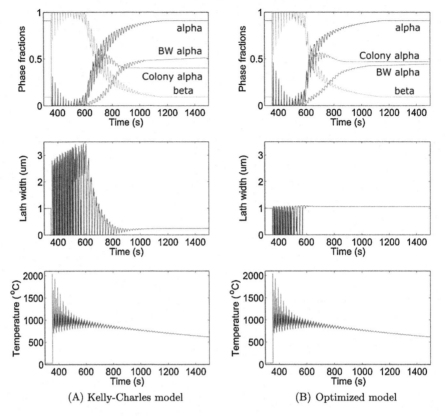

FIGURE 7.11 Phase fractions, lath width, and temperature versus time for both models at a single point in the middle of the 3-bead leg.

is very sensitive to composition, so the properties listed in Table 7.1 may not be appropriate for all alloys within the acceptable range of Ti-6Al-4V chemical composition.

Predicted colony-α phase fractions after the part has cooled are shown in Figure 7.12. The total α and β phase fractions return to their room temperature equilibrium values after cooling, $f_\alpha = 0.91$ and $f_\beta = 0.09$. It can be seen that the 4 s dwell time between layers is expected to result in less colony-α and finer laths (Figure 7.9) compared to 0 s dwell. Using Figure 7.12A as an example, several observations can be drawn in parallel with what Kelly found:

1. The large thermal mass of the substrate has an effect on the lowest layers, in this case, roughly layers 1–32 (0–5.7 mm above substrate)
2. The middle *characteristic layers* all have essentially the same microstructure, layers 33–235 (5.7–42.0 mm above substrate)
3. The top layers have not yet undergone as much thermal cycling as the characteristic layers and show differing amounts of colony alpha, layers 236–286 (42.0–51.2 mm above substrate)

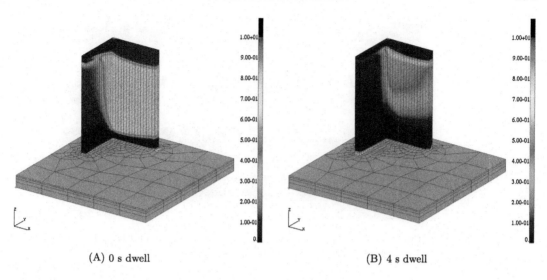

(A) 0 s dwell   (B) 4 s dwell

FIGURE 7.12 Colony-α phase fraction after the part has cooled.

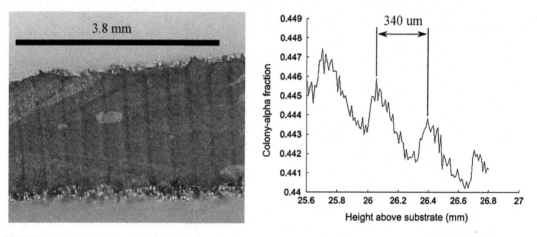

FIGURE 7.13 The reheating bands, as observed in a cross-section macrograph (left) and in the simulation results (right), for the 1-bead leg of the 0 s dwell part.

The phenomenon of characteristic layers in AM builds was observed previously by Kelly [21]. The morphology of these layers covers the majority of the finished part. If more layers are deposited, then the top layers would undergo additional thermal cycling and eventually transform to this characteristic morphology. For the 4 s dwell part, the characteristic region is smaller and more varied than in the 0 s dwell part.

Plotting the colony-α fraction against height for several layers reveals that the *reheating bands*, sometimes referred to as "layer bands," are reproduced by Kelly's model, as shown in Figure 7.13. These reheating bands occur when points along the height of the build are subjected to periodically varying thermal histories, resulting in a periodically varying mi-

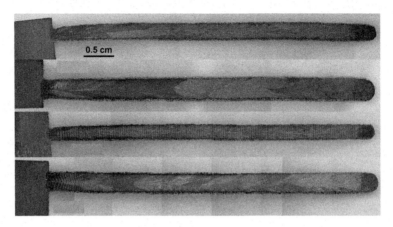

FIGURE 7.14 Cross-sections of both legs of both parts. From top to bottom: 1-bead leg 0 s dwell, 3-bead leg 0 s dwell, 1-bead leg 4 s dwell, and 3-bead leg 4 s dwell.

crostructures. The band spacing in the simulation matches that seen in optical micrographs, roughly corresponding to a bandwidth of two layers. The reheating bands are two layers thick rather than one because odd layers are deposited in the opposite direction as even layers, as shown in Figure 7.5. This change in deposition direction causes the thermal history of odd layers to be distinct from even layers. As shown in Figure 7.14, the bands appear in all cross-sections, but they are most evident for the 1-bead leg of the 4 s dwell part. While the 1-bead legs show flat layer bands, the 3-bead legs display bands with a scalloped shape. Note that the thermal simulation only has a spatial resolution of 180 μm, or one layer. Thus, the nodal temperatures had to be linearly interpolated between layers before being put into the microstructure model to obtain the results in Figure 7.13.

The prediction of α lath width allows for the prediction of mechanical properties and does not require destructive testing, unlike cross-sectioning, hardness indentation, or tensile testing. Lath width is closely linked to hardness, with thicker laths generally correlating to lower hardness [46]. Such a correlation, along with this microstructure model, could aid in part certification. By performing an in-situ measurement of temperatures during the process, e.g. with thermocouples or infrared cameras, final microstructures and mechanical properties could be calculated. Further, mechanical models of the residual stress and distortion that occur in AM processes could be improved. By linking the evolving microstructure to mechanical properties, a more realistic coupling of thermo-microstructural-mechanical models can be achieved.

7.5 CONCLUSIONS

Kelly's model for phase fraction and morphology evolution, and Charles's model for α lath width have been implemented for DED of Ti-6Al-4V. The accuracy of this model has been improved by using ANMS to obtain better estimates of material properties, reducing error by a factor of 3. The model has been validated with experimental measurements of lath width and has been shown to reproduce the observed reheating bands characterized by variations in colony and basketweave α fraction. While this work focuses on AM processes, the presented technique could be applied in future work to any process that involves repeated thermal cycles. A limitation of the current research is that only post-process measurements of lath width are used to validate the model. Future work could improve the validation by making in-situ or post-process measurements using X-ray or TEM techniques to also investigate phase fractions. Another limitation of this work is the weak coupling, i.e. the microstructure model is not influenced by the mechanical phenomena of the process. This limitation is not problematic for AM processes, but it may introduce errors for large-deformation processes, such as forging. Future work could expand the microstructure model by considering this coupling, by using measured temperatures instead of thermal simulations, and by correlating lath width more closely to hardness and other mechanical properties [47].

References

[1] Cai J, Li F, Liu T, Chen B, He M. Constitutive equations for elevated temperature flow stress of Ti-6Al-4V alloy considering the effect of strain. Mater Des 2011;32(3):1144–51.

[2] Majorell A, Srivatsa S, Picu R. Mechanical behavior of Ti-6Al-4V at high and moderate temperatures – Part I: experimental results. Mater Sci Eng A 2002;326(2):297–305.

[3] Tiley J, Searles T, Lee E, Kar S, Banerjee R, Russ J, et al. Quantification of microstructural features in α/β titanium alloys. Mater Sci Eng A 2004;372(1):191–8.

[4] Searles T, Tiley J, Tanner A, Williams R, Rollins B, Lee E, et al. Rapid characterization of titanium microstructural features for specific modelling of mechanical properties. Meas Sci Technol 2005;16(1):60.

[5] Costa L, Vilar R, Reti T, Deus A. Rapid tooling by laser powder deposition: process simulation using finite element analysis. Acta Mater 2005;53(14):3987–99.

[6] Kar S, Searles T, Lee E, Viswanathan G, Fraser H, Tiley J, et al. Modeling the tensile properties in β-processed α/β Ti alloys. Metall Mater Trans A 2006;37(3):559–66.

[7] Avrami M. Kinetics of phase change. I General theory. J Chem Phys 1939;7(12):1103–12.

[8] Avrami M. Kinetics of phase change. II Transformation-time relations for random distribution of nuclei. J Chem Phys 1940;8(2):212–24.

[9] Avrami M. Granulation, phase change, and microstructure kinetics of phase change. III. J Chem Phys 1941;9(2):177–84.

[10] Johnson WA, Mehl RF. Reaction kinetics in processes of nucleation and growth. Trans AIME 1939;135(8):396–415.

[11] Kolmogorov AN. On the statistical theory of the crystallization of metals. Bull Acad Sci USSR, Math Ser 1937;1:355–9.

[12] Malinov S, Markovsky P, Sha W, Guo Z. Resistivity study and computer modelling of the isothermal transformation kinetics of Ti-6Al-4V and Ti-6Al-2Sn-4Zr-2Mo-0.08 Si alloys. J Alloys Compd 2001;314(1):181–92.

[13] Mudge RP, Wald NR. Laser engineered net shaping advances additive manufacturing and repair. Weld J 2007;86(1):44–8.

[14] Melchels FP, Feijen J, Grijpma DW. A review on stereolithography and its applications in biomedical engineering. Biomaterials 2010;31(24):6121–30.

[15] Kobryn P, Semiatin S. The laser additive manufacture of Ti-6Al-4V. JOM 2001;53(9):40–2.

[16] Mahesh M, Wong Y, Fuh J, Loh H. Benchmarking for comparative evaluation of RP systems and processes. Rapid Prototyping J 2004;10(2):123–35.

[17] Sheng W, Xi N, Chen H, Chen Y, Song M. Surface partitioning in automated CAD-guided tool planning for additive manufacturing. In: Intelligent robots and systems, 2003. (IROS 2003). 2003 IEEE/RSJ international conference on proceedings, vol. 2. IEEE; 2003. p. 2072–7.

[18] Galantucci L, Lavecchia F, Percoco G. Experimental study aiming to enhance the surface finish of fused deposition modeled parts. CIRP Ann-Manuf Technol 2009;58(1):189–92.

[19] Huang Y, Leu MC, Mazumder J, Donmez A. Additive manufacturing: current state, future potential, gaps and needs, and recommendations. J Manuf Sci Eng 2015;137(1):014001.

[20] Panhalkar N, Paul R, Anand S. Increasing part accuracy in additive manufacturing processes using a kd tree based clustered adaptive layering. J Manuf Sci Eng 2014;136(6):061017.

[21] Kelly SM. Thermal and microstructure modeling of metal deposition processes with application to Ti-6Al-4V. Virginia Tech., 2004.

[22] Paul S, Gupta I, Sing RK. Characterization and modeling of microscale preplaced powder cladding via fiber laser. J Manuf Sci Eng 2015;137(3):031019.

[23] Huang Q, Nouri H, Xu K, Chen Y, Sosina S, Dasgupta T. Statistical predictive modeling and compensation of geometric deviations of three-dimensional printed products. J Manuf Sci Eng 2014;136(6):061008.

[24] Cheng B, Price S, Lydon J, Cooper K, Chou K. On process temperature in powder-bed electron beam additive manufacturing: model development and validation. J Manuf Sci Eng 2014;136(6):061018.

[25] Kelly S, Kampe S. Microstructural evolution in laser-deposited multilayer Ti-6Al-4V builds: part I. Microstructural characterization. Metall Trans A 2004;35(6):1861–7.

[26] Kelly S, Kampe S. Microstructural evolution in laser-deposited multilayer Ti-6Al-4V builds: part II. Thermal modeling. Metall Trans A 2004;35(6):1869–79.

[27] Christian JW. The theory of transformations in metals and alloys (part I + II). Newnes; 2002.

[28] Kriczky DA, Irwin J, Reutzel EW, Michaleris P, Nassar AR, Craig J. 3D spatial reconstruction of thermal characteristics in directed energy deposition through optical thermal imaging. J Mater Process Technol 2015;221:172–86.

[29] Fan Y, Cheng P, Yao Y, Yang Z, Egland K. Effect of phase transformations on laser forming of Ti-6Al-4V alloy. J Appl Phys 2005;98(1):013518.

[30] Crespo A. Modelling of heat transfer and phase transformations in the rapid manufacturing of titanium components. INTECH Open Access Publisher; 2011.

[31] Charles C. Modelling microstructure evolution of weld deposited Ti-6Al-4V. Luleå University of Technology; 2008.

[32] Charles C, Järvstråt N. Modelling Ti-6Al-4V microstructure by evolution laws implemented as finite element subroutines: application to TIG metal deposition. In: Proceedings of the 8th international conference "Trends in welding research"; 2009. p. 477–85.

[33] Murgau CC, Pederson R, Lindgren L. A model for Ti-6Al-4V microstructure evolution for arbitrary temperature changes. Model Simul Mater Sci Eng 2012;20(5):055006.

[34] Carroll BE, Palmer TA, Beese AM. Anisotropic tensile behavior of Ti-6Al-4V components fabricated with directed energy deposition additive manufacturing. Acta Mater 2015;87:309–20.

[35] Husain A, Sehgal D, Pandey R. An inverse finite element procedure for the determination of constitutive tensile behavior of materials using miniature specimen. Comput Mater Sci 2004;31(1):84–92.

[36] Hess R, Wang S, Gao C. Generalized technique for inverse simulation applied to aircraft maneuvers. J Guid Control Dyn 1991;14(5):920–6.

[37] Calvello M, Finno RJ. Selecting parameters to optimize in model calibration by inverse analysis. Comput Geotech 2004;31(5):410–24.

[38] Ahmed T, Rack H. Phase transformations during cooling in $\alpha + \beta$ titanium alloys. Mater Sci Eng A 1998;243(1):206–11.

[39] Gil F, Ginebra M, Manero J, Planell J. Formation of α-Widmanstätten structure: effects of grain size and cooling rate on the Widmanstätten morphologies and on the mechanical properties in Ti-6Al-4V alloy. J Alloys Compd 2001;329(1):142–52.

[40] Elmer J, Palmer T, Babu S, Zhang W, DebRoy T. Phase transformation dynamics during welding of Ti-6Al-4V. J Appl Phys 2004;95(12):8327–39.

[41] Nelder JA, Mead R. A simplex method for function minimization. Comput J 1965;7(4):308–13.

[42] Gao F, Han L. Implementing the Nelder-Mead simplex algorithm with adaptive parameters. Comput Optim Appl 2012;51(1):259–77.

[43] Denlinger ER, Heigel JC, Michaleris P. Residual stress and distortion modeling of electron beam direct manufacturing Ti-6Al-4V. Proc Inst Mech Eng, B J Eng Manuf 2014:0954405414539494.

[44] Gouge MF, Heigel JC, Michaleris P, Palmer TA. Modeling forced convection in the thermal simulation of laser cladding processes. Int J Adv Manuf Technol 2015:1–14.

[45] Heigel J, Michaleris P, Reutzel E. Thermo-mechanical model development and validation of directed energy deposition additive manufacturing of Ti-6Al-4V. Add Manuf 2014.

[46] Chlebus E, Kuźnicka B, Kurzynowski T, Dybała B. Microstructure and mechanical behaviour of Ti–6Al–7Nb alloy produced by selective laser melting. Mater Charact 2011;62(5):488–95.

[47] Tayon WA, Shenoy RN, Redding MR, Bird RK, Hafley RA. Correlation between microstructure and mechanical properties in an inconel 718 deposit produced via electron beam freeform fabrication. J Manuf Sci Eng 2014;136(6):061005.

8

Thermo-Mechanical Modeling of Thin Wall Builds using Powder Fed Directed Energy Deposition*

Jarred C. Heigel

The Pennsylvania State University,[a] University Park, PA, USA

8.1 INTRODUCTION

Directed Energy Deposition (DED) [1] is an additive manufacturing (AM) process that creates parts through the layer-by-layer addition of material. DED uses a high intensity energy source, such as a laser, to create a melt pool into which metal powder or wire is injected. This differs from the more widely used powder bed fusion (PBF), which use a high energy source to melt pre-placed layers of powder. Each of these processes use a heat source that follows a pattern to solidify each layer, progressively building the part. The resulting complex thermal history influences the residual stress and distortion of the final part. Thermo-mechanical finite element analysis (FEA) can be used to model these effects, enabling high quality parts to be produced.

FEA modeling of DED is inspired by weld modeling, since it is a similar process that has been studied extensively [2–4]. Although many of the weld modeling studies are directly applicable to DED modeling efforts, the convection models used are not applicable. Some weld studies have achieved useful results by neglecting convection while others have applied free convection uniformly on all exposed surfaces. These approaches lead to small errors in weld modeling because of the small amount of filler material relative to the substrate, which allows most of the heat to be conducted away from the bead into the parts being joined. In contrast, filler material makes up the majority of a part built using DED, resulting in longer processing times and higher temperatures that allow for a greater amount of heat

---

* Portions of this text are reproduced from Heigel, Michaleris, Reutzel, 2015, "Thermo-mechanical model development and validation of directed energy deposition additive manufacturing of Ti-6Al-4V," Additive Manufacturing, V5. with permission from Elsevier.

[a] The work was performed while the author was a graudate research assistant at the Pennsylvania State University. The author is currently at the National Institute of Standards and Technology, Gaithersburg MD, 20899.

loss through convection. Consequently, greater errors can occur from inaccurate convection models in DED simulations. Complex convection models are required because of the inert gas jets often used to protect the laser optics, to shield the molten material from oxidation, and to aid in delivering powder to the melt pool. The heat transfer literature demonstrates that these types of jets create localized forced convection that is influenced by a variety of factors [5–7].

The literature shows inconsistent implementation of convection in DED models. Heat loss due to convection is assumed negligible and excluded in some models [8–11]. Convection is incorporated in other models by assuming it is uniformly distributed over all surfaces and equal to free convection in air ($10 \, W \, m^{-2} \, K^{-2}$) [12–20] while others have applied a higher uniform convection [21,22], presumably to account for the greater amount of surface convection caused by the inert gas jets. Some researchers have considered the complexity of forced convection when modeling DED. Ghosh and Choi used the empirical equation defined by Gardon [23] to account for the forced convection [24]. Zekovic and co-workers included forced convection when modeling a thin wall deposition by using Computational Fluid Dynamics (CFD) to calculate the convection acting on the surface [25]. However, there was no experimental effort to validate the CFD results for their process. Furthermore, no work has been found in the literature that develops a measurement-based forced convection model.

This chapter presents the development of a thermo-mechanical model for the deposition of thin walled structures using a DED process. The intent is to provide greater insight into the initial experimental and numerical investigation that led to the model initially presented in [26]. Although the simulation approach has been discussed in detail in earlier chapters, it is presented again here in the context of the relevant considerations for DED processing and their differences with other AM processes. Next, the specific DED process used in this study is presented. Following this, the different deposition cases and the in-situ measurement setup are presented. Only after the DED process and the deposition parameters are understood can the model inputs and the numerical implementation of the processes be understood. The sections that follow present the inputs that are found in the literature and describe the initial investigation into the heat transfer input variables that could not be found in the literature. Ultimately, this investigation led to the realization that a measurement-based model of surface convection was required to develop a quality model. This chapter concludes by comparing the validated model results to the in-situ measurements of the actual deposition and the possible impact from assumptions used to construct the model is discussed.

8.2 DED SIMULATION

The FEA mathematics presented in Chapter 2 are used to predict the thermal history and mechanical response of the DED process discussed in this chapter. Although these equations are also used to model other metal AM process, such as electron beam DED (Chapter 9), laser cladding with a variety of feedstocks [27], and powder bed fusion processes (Chapters 12–15), it is the definition of the material (ρ, C_p, k, E, σ_y, and α), the heat input (P, η, f, a, b, and c), and the surface heat loss (ε, h, and β) that distinguishes the model of each process. However,

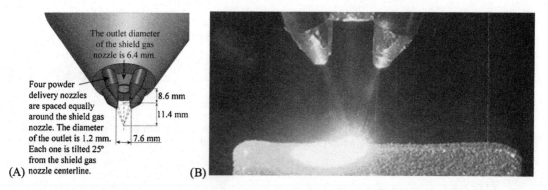

FIGURE 8.1 Overview of the DED system modeled in this chapter. (A) the deposition head (Reprinted from [26], with permission for Elsevier). (B) Deposition of a 38 mm long thin wall.

before those variables can be defined for the DED process of interest in this study, the process must first be understood.

8.3 DED PROCESS MEASUREMENT SETUP AND TEST CASES

Single track thin walls of Ti-6Al-4V are deposited using an Optomec® LENS MR-7 system with a 500 W IPG Photonics fiber laser. The deposition occurs in a chamber with an argon atmosphere that has an oxygen content of less than 15 parts per million. A 30 L/min argon jet is used to supply argon to the chamber, to protect the laser optics, and to shield the melt pool. The Ti-6Al-4V powder delivered to the melt pool is assisted by four argon jets that have a combined flow rate of 4 L/min. These four jets exit nozzles positioned around the main nozzle and aimed at the melt pool, as shown in Figure 8.1A. The powder has been sieved so that only particles with diameters between 44 μm and 149 μm are delivered at a rate of 3.0 g/min. Figure 8.1B shows the powder being delivered to the melt pool during the deposition of one of the thin walls created in this study.

8.3.1 Deposition Cases

In order to develop and validate the model, a variety of straight and relatively tall thin walls are built while the thermal and mechanical responses are measured in-situ. Although AM technologies can be used to build parts with very complex geometries, straight walls simplify the build, making it easier to interpret measurement results and allow the FEA analysis to be performed using a half-symmetry model, significantly reducing the computational load.

The model is validated using three different depositions, as shown in Figure 8.2. Each case builds a wall that is designed to be 38.1 mm long, 12.7 mm tall, and 3 mm wide. These cases produce different thermal and mechanical results that are used to validate the model:

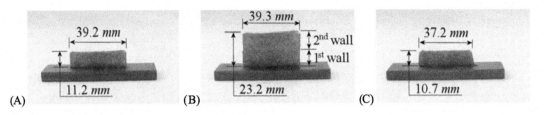

FIGURE 8.2 Images of the deposited thin walls, reprinted from [26] with permission from Elsevier. (A) Single wall with no dwell between layers. (B) Double wall with no dwell between layers. (C) Single wall with a 20 s dwell between layers.

TABLE 8.1 The test cases and process conditions used. Reprinted from [26] with permission from Elsevier.

Case	1	2	3
Measured laser power (W)	410	415	415
Travel speed (mm/s)	8.5	8.5	8.5
Powder delivery rate (g/min)	3.0	3.0	3.0
Additional dwell between layers (s)	0	0	20
Total wall height (mm)	11.2	23.2	10.7
Measured wall length (mm)	39.2	39.3	37.2
Measured wall width (mm)	3.0	3.1	2.2

1. A single wall built using 62 layers, each one track wide, that are deposited without any dwell between layers onto a 76.2 mm long, 25.4 mm wide and 6.4 mm thick Ti-6Al-4V substrate.
2. A 2nd 62 layer wall is deposited on top of the wall built in Case 1 without any dwell between each layer. This results in a final deposition that is a total of 124 layers, hereafter referred to as the double wall. This deposition increases the area over which the forced convection acts.
3. A 62 layer wall is deposited onto a substrate with a 20 s dwell between each layer. This generates lower temperatures compared to the deposition with no dwell (Case 1).

In each case layers are deposited in alternating directions, such that during odd layers material is deposited left to right and during even layers material is dposited right to left. Table 8.1 presents the process conditions used in each case. In all depositions the nominal power is 500 W; however, power measurements made using a Macken P500 power probe (with an accuracy of ±25 W) before each deposition show that the actual power being supplied by the laser is between 410 W and 415 W.

8.3.2 Measurement Setup

While in-situ measurement of temperature on the substrate is relatively straightforward using thermocouples pre-attached to the part, measuring the mechanical response in-situ is more challenging. Residual stress accumulates during the build, but is not easily measured. However, the resulting strain and distortion can be measured if the part is not constrained during the build. In this study, as in many other presented in this book, in-situ deflection

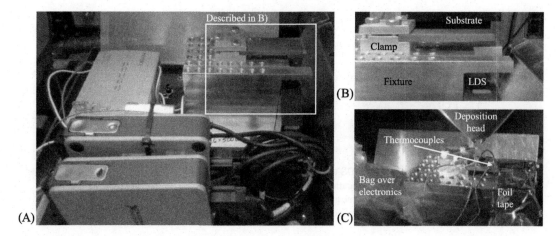

FIGURE 8.3 The measurement setup. (A) The mixture and all accompanying electronics in the build chamber, (B) the substrate in the clamp above the LDS (reprinted from [26] with permission from Elsevier), and (C) the setup just before a deposition.

of the substrate is measured by clamping one of its ends into a fixture, cantilevering the free end over a Laser Displacement Sensor (LDS). The LDS used in this study is a Keyence LK-031, which has a measurement accuracy of ±1 μm. It measures the vertical distance to a point on the bottom surface of the substrate. In-situ temperature is measured at the center of the bottom surface of the substrate using an Omega GG-K-30 type K thermocouple. The thermocouple has a measurement uncertainty of 2.2 °C or 0.75%, whichever is larger. While all of the electronics to record these signals are placed inside the build chamber (Figure 8.3A), they are protected from powder during the build using a high temperature plastic bag, as shown in Figure 8.3C.

8.4 RESULTS FROM THE IN-SITU MEASUREMENTS

Figure 8.4 presents the deposited wall and the measured temperature and deflection of Case 1. The temperature measured at the bottom center of the substrate using the thermocouple is represented by the black curve. The gray curve represents the deflection near the free end of the substrate measured using the LDS. As the temperature rapidly rises at the beginning of the process, it fluctuates more during the beginning of the build since there is less material between the melt pool and the thermocouple on the bottom of the substrate. After approximately 150 s of processing time the temperature on the bottom of the substrate begins to decrease despite the continual addition of heat from the laser on top of the wall to build it up. This suggests that the build is sufficiently tall to enable the surface heat loss through convection and radiation to dissipate much of the heat supplied by the laser before it can be conducted to the bottom of the substrate. Therefore, this case will provide a sufficient test of the surface heat flux models implemented in the FEA. Regarding the LDS measurement, the

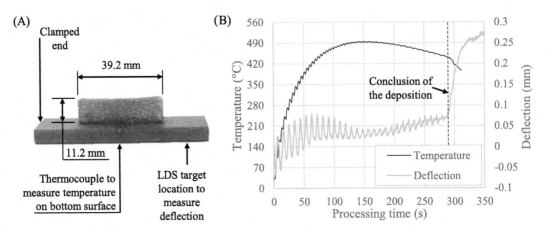

FIGURE 8.4   Case 1 (A) deposited wall and (B) measurement results.

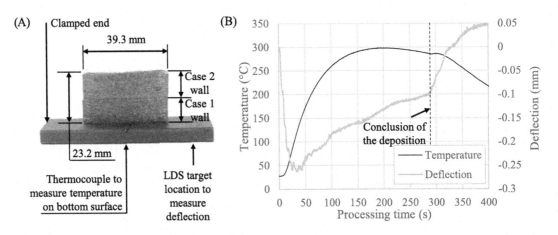

FIGURE 8.5   Case 2 (A) deposited wall and (B) measurement results.

distortion follows the trend observed by Denlinger and his colleagues, who demonstrated that when processing Ti-6Al-4V, sufficiently high temperatures causes the material to anneal during the process, limiting the amount of distortion that is accumulated [28]. More detail regarding that phenomena can be found in Chapters 9–11.

The second wall deposited in Case 2 and the measurement results are presented in Figure 8.5. Since the 62 layer wall in Case 2 is deposited on top of the Case 1 wall, it provides additional insight into the process that could not be obtained in Case 1. First, as expected, the thermocouple on the substrate measures an increasing temperature that peaks before the end of the process, after that it slowly decreases till the end of the test. Once again, this is a result of the increasing size of the build which further removes the melt pool from the thermocouple location and provides a greater area through which heat can be removed by convection and radiation. However, this effect is more pronounced due to the existing wall from Case 1,

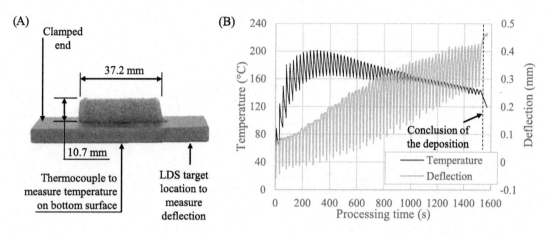

FIGURE 8.6 Case 3 (A) deposited wall and (B) measurement results.

reducing the peak temperature compared to the previous case. Second, the distortion is drastically different than in the previous cases. Here, the plate steadily deflects downward as the entire wall is heated and expands as new material is added. After a short time, the part begins to deflect upwards as the temperature evens out in the part and concludes with little additional upward deflection as compared to that with which it began.

Figure 8.6 presents the results for Case 3, which imposes a 20 s dwell between layers. This dwell significantly affects both the temperature and distortion compared to Case 1, which does not impose the dwell. For instance, peak temperature is lower (200 °C compared to 490 °C) and occurs sooner into the build. These differences occur because although the total heat input by the laser is approximately the same, the dwells provide more time for the heat to escape through convection and radiation. The lower temperature experienced during this case affects the substrate deflection, which oscillates up and down in response to the movement of the laser, but steadily increases during the whole test. Once again, this trend is consistent with the prior work by Denlinger et al. [28].

Each of these cases provide unique insight into the process of building thin wall structures and are useful in developing and validating the FEA model. Case 1 is used to develop the FEA model of the process since it deposits a significant amount of material over a long enough processing time to allow all modes of heat transfer to be tested. Cases 2 and 3 will not be used to develop the model, but will be used to validate the accuracy of the model, specifically in regard to balancing the energy input from the laser with the energy lost through convection and radiation.

8.5 NUMERICAL IMPLEMENTATION

The FEA analysis is performed using Netfabb Simulation using the hybrid quiet-inactive element activation method to simulate the deposition of material during the DED process [29]. The material properties for Ti-6Al-4V used in this study can be found in the appendix.

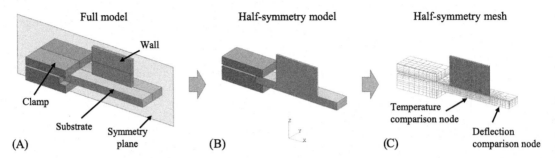

FIGURE 8.7   (A) A solid model of the substrate, wall, and fixture clamp, (B) the half-symmetry solid model, and (C) the 3D finite element mesh. Reprinted from [26] with permission from Elsevier.

8.5.1 Finite Element Mesh

Figure 8.7 shows the 3-dimensional half-symmetry mesh that is used in the thermal and mechanical simulations of Cases 1 through 3. There are a total of 23295 nodes and 15627 elements used to define the fixture's aluminum clamp that holds one end of the substrate, the titanium alloy substrate, and the wall that is deposited during the process. The aluminum clamp is included in this model since, as discussed in Chapter 4, the heat transfer from the substrate into the fixture can significantly affect the accuracy of the thermal model and the subsequent mechanical analyses. However, in the case of cantilevered workpieces in DED processes, Gouge et al. found that, when compared to heat transfer through convection and radiation, the heat transfer through conduction into the fixture is negligible [30]. The substrate, which is mechanically constrained so that the clamped end remains fixed and the free end is allowed to deform, is composed of elements that are fine near the initial laser pass and coarse near the edges. This reduces the number of equations and the computational time. The majority of the elements, 11904 to be exact, are used to define the deposited wall.

Errors between measurement and simulation results are calculated by comparing single instances in time or by calculating the percent error over the entire deposition time:

$$\% \text{ Error} = \frac{100 \sum_{i=1}^{n} \left| \frac{(T_{\text{meas}})_i - (T_{\text{sim}})_i}{(T_{\text{meas}})_i} \right|}{n} \tag{8.1}$$

where n is the total number of simulated time increments during the deposition, i is the current time increment, T_{sim} is the simulated temperature at the node that corresponds to the location of the thermocouple, and T_{meas} is the measured temperature.

8.5.2 Determination of the Heat Source and Surface Loss Variables

After the mesh and the material properties are defined, the remaining variables that need to be determined relate to the heat source and surface heat loss. Concerning the heat source, the absorption efficiency, η, is determined to be 0.45 through inverse simulation based on the first few deposition tracks. This will be discussed in more detail. The emissivity, ε, of Ti-6Al-4V in an electron beam DED processes was determined by Denlinger and his col-

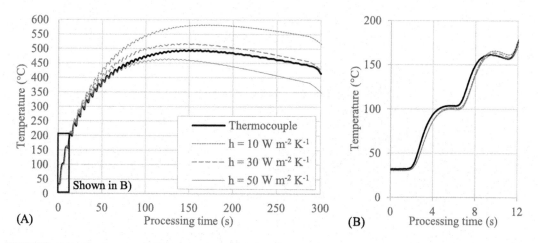

FIGURE 8.8 The impact of different values of the coefficient of convection applied uniformly to all part surface. (A) The temperature measured by the thermocouple compared to simulation results using $10\ \mathrm{W\,m^{-2}\,K^{-1}}$, $30\ \mathrm{W\,m^{-2}\,K^{-1}}$, and $50\ \mathrm{W\,m^{-2}\,K^{-1}}$. (B) Passes 1 through 3 and (C) passes 20 through 22.

leagues to be 0.54 [31]. The remaining unknown for the DED processes being modeled in this study is the coefficient of convection, h.

Since the majority of literature on AM modeling assumes a simple convection model [12–22], this study begins with the same assumption. Figure 8.8 presents a few example results from this preliminary calibration using Case 1, the single wall deposition with no dwell between layers assuming an absorption efficiency of $\eta = 0.45$ and values of h that range from 10 to $50\ \mathrm{W\,m^{-2}\,K^{-1}}$. These results show that although the different values of h significantly impact the temperature, they have no effect during the first few deposition tracks, as shown in Figure 8.8B, because the conduction into the workpiece dominates the energy balance. However, as the workpiece begins to heat up, the effects of surface heat loss become significant. Therefore, these initial tracks are relatively insensitive to surface heat loss and can be used to validate the chosen absorption efficiency of $\eta = 0.45$.

Figure 8.8 also proves that the assumption of natural convection or higher uniform values of h is incapable of accurately predicting the temperature during the entire build. First, the value of $h = 10\ \mathrm{W\,m^{-2}\,K^{-1}}$ clearly does not evacuate enough heat from the part. Elevated values of h ($30\ \mathrm{W\,m^2\,K^{-1}}$ and $50\ \mathrm{W\,m^2\,K^{-1}}$) demonstrate that convection can significantly reduce the temperature, but the assumed uniform value cannot accurately predict the shape of the temperature profile. For instance, a value of $h = 50\ \mathrm{W\,m^2\,K^{-1}}$ allows the simulation to match the measurement for the first 15 tracks (up to 68 s), but afterwards the temperature rapidly decreases below the measurement and, in fact, the temperature predicted using a value of $h = 30\ \mathrm{W\,m^2\,K^{-1}}$ is closer to predicting the measurement by the end of the 62nd track (294 s). Without a large number of depositions, the inadequacy of a uniform value of h on all surfaces would not be realized.

A more complex convection model is justified considering the argon jets emitted from the deposition head to shield the laser optics and to deliver powder to the melt pool and that these jets are constantly focused at the hottest location of the part. Therefore, a simpli-

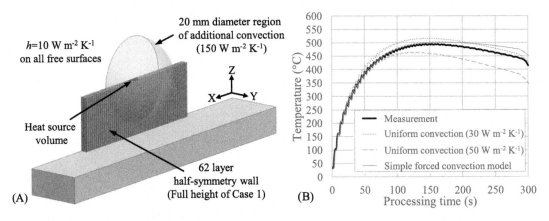

FIGURE 8.9 A simplified model of forced convection from the argon jets. (A) Illustration of the model, (B) simulation results.

fied model of this forced convection is developed to explore its possible impact. Figure 8.9A illustrates this model, which applies a uniform convection of $10\ \mathrm{W\,m^2\,K^{-1}}$ on all free surfaces, but increases h by $150\ \mathrm{W\,m^2\,K^{-1}}$ within a 20 mm diameter sphere centered at the heat source. Figure 8.9B presents these results compared to a uniform values of $h = 30\ \mathrm{W\,m^2\,K^{-1}}$ and $h = 50\ \mathrm{W\,m^2\,K^{-1}}$. Increasing the convection around the heat source changes the shape of the temperature profile. Although an exhaustive investigation could be undertaken to find values of h inside and outside of that sphere, it would still be possible that there was an unaccounted-for effect, such as changing laser absorption efficiency that would be masked by such a study. Therefore, a measurement-based convection model is required.

A convection model is developed based on measurements acquired using hot-film sensors and constant voltage anemometry. Sensors are mounted to a thin plate which is located in the X–Y plane to measure the effects of the argon jets impinging onto the substrate, or placed in the X–Z plane to measure the forced convection from the jets being bisected by a vertical wall. Both the measurement process and the methodology to calculated h are detailed in [26,32]. The empirically derived model to describe the convection acting on the deposited vertical wall (h_{wall}) and on the substrate (h_{surface}) is:

$$h_{\mathrm{wall}} = (-2.7z + 37.2)e^{-((0.107r)^{2.7})} + 25\ \mathrm{W\,m^{-2}\,K^{-1}} \tag{8.2}$$

$$h_{\mathrm{surface}} = (-5.1z + 70.7)e^{-((0.031r)^{1.4})} + 30\ \mathrm{W\,m^{-2}\,K^{-1}} \tag{8.3}$$

where r is the distance from the deposition head centerline to the point of interest and z is the vertical distance from the top edge of the wall to the point of interest. A more detailed description of these models and their development is provided in [26].

The new empirically derived convection model is demonstrated in Figure 8.10, which compares the simulated and measured temperature and illustrates the convection acting on the part at various stages during the build. The measurement-based convection model greatly improves the simulation results, making them and the measured temperature curves nearly

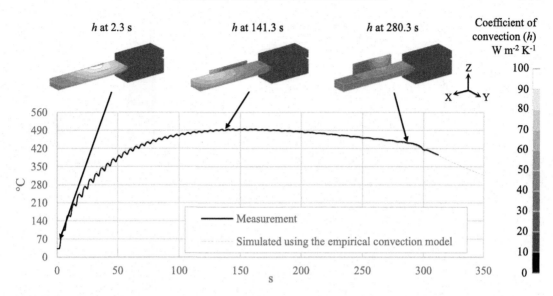

FIGURE 8.10 Comparison of the simulation results using the empirical convection model to the measured temperature.

indistinguishable. Therefore, the thermal model is now developed and can be applied to each case to predict the temperature and to be used as an input to the mechanical model.

8.6 THERMO-MECHANICAL MODELING RESULTS

Figure 8.11 compares the thermal and mechanical results for each case to the measurements. In subfigures A, B, and C, the measured temperature and deflection are plotted using solid black and gray lines, respectively, while the corresponding simulation results are plotted with dashed lines. The temperature and distortion history plots are accompanied by example temperature distributions from a single time step taken at the approximate half-way point of the build (B, C, and D). A circle is added to the temperature and distortion history plots to indicate where in time the mesh results are taken.

The results for Case 1 are shown in Figure 8.11A and B. As discussed earlier and initially shown in Figure 8.10, the simulated temperature on the bottom of the substrate in Case 1 is an excellent match with the measured temperature, having a percent error of only 2.4%. The mechanical simulation predicts the correct deflection trend, where the substrate oscillates up and down for most of the deposition, but does not begin to significantly curl upwards until the conclusion of the deposition. However, despite the quality of the thermal simulation, the magnitude of the simulated distortion is approximately 50 μm less than the measurement.

Figures 8.11C and D present the results for the deposition of the second wall in Case 2. The temperature does not follow the measurement as closely as in Case 1. Compared to the measurement, the simulated temperature rises more rapidly, reaches its maximum value

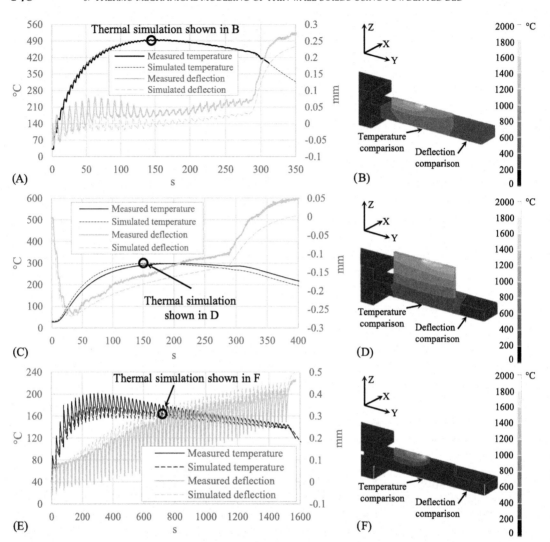

FIGURE 8.11 Thermo-mechanical simulation results for the single wall deposition with no dwell (Case 1, A and B), the double wall with no dwell between layers (Case 2, C and D), and the single wall with a 20 s dwell between layers (Case 3, E and F). (B), (D), and (F) are reprinted from [26] with permission from Elsevier.

sooner, and cools more during the remainder of the deposition, although the percent error is still quite good at 4.1%. The increased thermal simulation error could be due to a number of factors. First, the temperature dependent material properties are lacking at temperatures above 870 °C and are consequently assumed to be constant in the current model. Figure 8.11D shows that a significant portion of the wall exceeds this temperature compared to Case 1 (Figure 8.11B). Therefore, Case 2 would be affected more by the lack of high-temperature material property data. A second explanation is that the higher wall temperatures and larger surface area in Case 2 are able to test the convection model more strenuously, and show that it can be improved.

Although the thermal simulation results are not as accurate for Case 2, the subsequent mechanical analysis captures the correct deflection trend. Here, the workpiece rapidly distorts downward as the wall heats up, then at about the same time as the simulation, begins to distort upward. Although the trend is once again captured, the simulation underpredicts the final distortion by about 40 μm.

The simulation results for Case 3 are presented in Figure 8.11E and F. Like the measurement, the temperature rises for approximately 300 s to a maximum value, after which it gradually decreases. During the entire process, the temperature fluctuates as the melt pool passes closer to the laser. However, despite the similar trend, the simulation predicts lower temperatures and results with an error of 10.4%. Since the part temperatures are much lower in this case than in Case 1 (Figure 8.11F vs. Figure 8.11B), the increased error in the thermal model is likely due to the forced convection model. It could be improved, possibly with a validated CFD model or more extensive empirical measurements.

The mechanical analysis is also able to capture the correct trend in Case 3, where although the part distortion fluctuates up and down in response to the laser position, it experiences a regular net increase in deflection during the entire test. Unlike Cases 1 and 2, the simulated distortion slightly exceeds that of the measurement by about 50 μm.

As discussed earlier, the difference in the mechanical response between Cases 1 and 3 is due to the temperature of the workpiece, and the simulations results presented in Figure 8.11B and F demonstrate this effect. At the mid-point of the deposition, a significant amount of material in Case 1 is above the stress relaxation temperature of 690 °C, preventing stress from accumulating in the part. In contrast, the dwell times in Case 3 keep the part cool enough that very little of the material is above the stress relaxation temperature; therefore, the stress is able to accumulate and the distortion increases steadily during the build process.

8.7 CONCLUSIONS

A model of the deposition of a thin wall structure using a powder fed DED process is presented. The model was developed using the in-situ measurements of temperature and distortion of a 62 layer, single track wide wall. During the development of the thermal simulation, simple models of the convection, similar to those commonly used in the literature, proved to be insufficient and could not capture the heating and cooling trends in the substrate. Consequently, a forced convection model was developed based on measurements of the convection resulting from the inert gas jets emitted by the deposition head. Incorporating the empirically-derived forced convection model resulted in a simulation error of 4.1%.

Two additional depositions were made to further test the model and demonstrate its accuracy. One deposition added a second 62 layer wall on top of the first, while the final deposition created a new 62 layer wall, but with 20 s dwells between each layer. While the thermal model was able to capture the proper trends, the error for each of these cases was greater. The mechanical analysis of each of the three cases was able to predict the correct trends, but the magnitude differed by about 50 μm.

The following conclusions can be made from this study:

1. The first few passes of a deposition can be used to calibrate the energy absorption efficiency, since conduction into the workpiece dominates the energy balance.
2. Models must be validated with relatively large builds. This allows all modes of heat transfer to become significant as the part heats up and provides enough time for error to become pronounced.
3. Uniform convection cannot be assumed when modeling DED processes if gas jets are emitted from the deposition head. Force convection must be used.

References

[1] Standard Terminology for Additive Manufacturing – General Principles – Terminology. ISO/ASTM, Standard 52900:2015(E).
[2] Lindgren L-E. Finite element modeling and simulation of welding. Part 1: increased complexity. J Therm Stresses 2001;24(2):141–92.
[3] Lindgren L-E. Finite element modeling and simulation of welding. Part 2: improved material modeling. J Therm Stresses 2001;24(3):195–231.
[4] Lindgren L-E. Finite element modeling and simulation of welding. Part 3: efficiency and integration. J Therm Stresses 2001;24(4):305–34.
[5] Perry KP. Heat transfer by convection from a hot gas jet to a plane surface. Proc Inst Mech Eng 1954;168(1):775–84.
[6] Gardon R, Akfirat JC. The role of turbulence in determining the heat-transfer characteristics of impinging jets. Int Commun Heat Mass Transf 1965;8(10):1261–72.
[7] O'Donovan TS, Murray DB. Jet impingement heat transfer–Part I: mean and root-mean-square heat transfer and velocity distributions. Int Commun Heat Mass Transf 2007;50(17):3291–301.
[8] Tikare V, Griffith M, Schlienger E, Smugeresky J. Simulation of coarsening during laser engineered net-shaping. Albuquerque, NM (United States): Sandia National Labs; 1997.
[9] Ensz M, Griffith M, Hofmeister W, Philliber JA, Smugeresky J, Wert M. Investigation of solidification in the laser engineered net shaping (LENS) process. Albuquerque, NM, and Livermore, CA: Sandia National Laboratories (SNL); 1999.
[10] Pinkerton A, Li L. The development of temperature fields and powder flow during laser direct metal deposition wall growth. Proc Inst Mech Eng, Part C, J Mech Eng Sci 2004;218(5):531–41.
[11] Vasinonta A, Beuth JL, Griffith M. Process maps for predicting residual stress and melt pool size in the laser-based fabrication of thin-walled structures. J Manuf Sci Eng 2007;129(1):101–9.
[12] Jendrzejewski R, Śliwiński G, Krawczuk M, Ostachowicz W. Temperature and stress fields induced during laser cladding. Comput Struct 2004;82(7):653–8.
[13] Jendrzejewski R, Kreja I, Śliwiński G. Temperature distribution in laser-clad multi-layers. Mater Sci Eng A 2004;379(1):313–20.
[14] Labudovic M, Hu D, Kovacevic R. A three dimensional model for direct laser metal powder deposition and rapid prototyping. J Mater Sci 2003;38(1):35–49.
[15] Kelly S, Kampe S. Microstructural evolution in laser-deposited multilayer Ti-6Al-4V builds: Part II. Thermal modeling. Metall Mater Trans A 2004;35(6):1869–79.
[16] Zheng B, Zhou Y, Smugeresky J, Schoenung J, Lavernia E. Thermal behavior and microstructural evolution during laser deposition with laser-engineered net shaping: Part I. Numerical calculations. Metall Mater Trans A 2008;39(9):2228–36.
[17] Pratt P, Felicelli S, Wang L, Hubbard C. Residual stress measurement of laser-engineered net shaping AISI 410 thin plates using neutron diffraction. Metall Mater Trans A 2008;39(13):3155–63.
[18] Wang L, Felicelli SD, Pratt P. Residual stresses in LENS-deposited AISI 410 stainless steel plates. Mater Sci Eng A 2008;496(1):234–41.
[19] He X, Yu G, Mazumder J. Temperature and composition profile during double-track laser cladding of H13 tool steel. J Phys D, Appl Phys 2009;43(1):15502.

[20] Lundbäck A, Lindgren L-E. Modelling of metal deposition. Finite Elem Anal Des 2011;47(10):1169–77.

[21] Hoadley A, Rappaz M, Zimmermann M. Heat-flow simulation of laser remelting with experimenting validation. Metall Trans B 1991;22(1):101–9.

[22] Dai K, Shaw L. Distortion minimization of laser-processed components through control of laser scanning patterns. Rapid Prototyping J 2002;8(5):270–6.

[23] Gardon R. Heat transfer between a flat plate and jets of air impinging on it. Int Dev Heat Transfer (ASME) 1962:454–60.

[24] Ghosh S, Choi J. Three-dimensional transient finite element analysis for residual stresses in the laser aided direct metal/material deposition process. J Laser Appl 2005;17(3):144–58.

[25] Zekovic S, Dwivedi R, Kovacevic R. Thermo-structural finite element analysis of direct laser metal deposited thin-walled structures, presented at the Proceedings SFF Symposium, Austin, TX, 2005.

[26] Heigel JC, Michaleris P, Reutzel EW. Thermo-mechanical model development and validation of directed energy deposition additive manufacturing of Ti-6Al-4V. Additive Manuf 2015;5:9–19.

[27] Heigel J, Gouge M, Michaleris P, Palmer T. Selection of powder or wire feedstock material for the laser cladding of Inconel® 625. J Mater Process Technol 2016;231:357–65.

[28] Denlinger ER, Heigel JC, Michaleris P, Palmer TA. Effect of inter-layer dwell time on distortion and residual stress in additive manufacturing of titanium and nickel alloys. J Mater Process Technol 2015;215:123–31.

[29] Michaleris P. Modeling metal deposition in heat transfer analyses of additive manufacturing processes. Finite Elem Anal Des 2014;86(1):51–60.

[30] Gouge M, Michaleris P, Palmer T. Fixturing effects in the thermal modeling of laser cladding. J Manuf Sci Eng 2017;139(1):11001.

[31] Denlinger ER, Heigel JC, Michaleris P. Residual stress and distortion modeling of electron beam direct manufacturing Ti-6Al-4V. Proc Inst Mech Eng, B J Eng Manuf 2014;229:1803–13.

[32] Heigel JC, Michaleris P, Palmer TA. Measurement of forced surface convection in directed energy deposition additive manufacturing. Proc Inst Mech Eng, B J Eng Manuf 2016;230(7):1295–308.

Residual Stress and Distortion Modeling of Electron Beam Direct Manufacturing Ti-6Al-4V*

Erik R. Denlinger

Product Development Group, Autodesk Inc., State College, PA, United States

9.1 INTRODUCTION

Additive Manufacturing (AM) has seen increased attention in recent years due to the ability of the process to produce near-net shape parts directly from CAD files without the retooling cost associated with casting or forging. The electron beam deposition process is of particular interest to the aerospace industry due to its ability to deposit large amounts of feedstock material at rapid rates. The process involves melting metal wire onto a substrate, and allowing for it to cool and form a fully dense geometry on a layer by layer basis. Unfortunately, large thermal gradients during the deposition process result in undesirable distortion and residual stress. Modifications to the build plan may reduce distortion and residual stress. To optimize the build plan without the expensive trial and error iterations an accurate predictive model is needed.

Numerical modeling for the prediction of temperature, distortion, and residual stress caused by the additive manufacturing process is similar to that of multi-pass welding. Weld modeling has been an active area of research for nearly 4 decades. Several weld models have been used to predict thermal and mechanical behavior [1–7]. Other modeling work has focused on material phase change during welding [8,9]. Lindgren has written detailed summaries on the development of weld model complexity [10], material modeling in welding [11], and improvements in computational efficiency for weld modeling [12]. In more recent weld model work, Michaleris et al. focused on predicting distortion modes caused by welding, as well as residual stress [13].

---

* This chapter is based upon the original work: Denlinger, Erik R., Jarred C. Heigel, and Pan Michaleris. "Residual stress and distortion modeling of electron beam direct manufacturing Ti-6Al-4V." Proceedings of the Institution of Mechanical Engineers, Part B: Journal of Engineering Manufacture 229.10 (2015): 1803–1813.

AM modeling adds significant computational cost when compared to multi-pass welding due to the increased amount of deposited material, passes, and process time. The addition of material into the simulation requires an element activation strategy [14]. Like weld models, AM models require accurate process parameters to yield acceptable results. The parameters of particular interest are absorption efficiency and surface emissivity, as these parameters determine the energy entering and exiting the system during the process. Yang et al. experimentally determined the absorption efficiency and surface emissivity of deposited Ti-6Al-4V using a laser assisted machining process [15]. Shen and Chou modeled the efficiency of a powder based electron beam system as 0.90 and assumed the emissivity of Ti-6Al-4V to be a constant 0.70, resulting in close agreement with experimental results [16]. No available literature provides values of efficiency and emissivity for a wire-fed electron beam system. Over the past decade significant work has been performed to model AM [17–20].

Some researchers have focused on material phase change caused by AM such as predicting the resulting microstructure of deposited Ti-6Al-4V [21–27]. Additional work has focused on predicting stress and distortion in materials that undergo phase transformation [28]. Ghosh and Choi concluded that simulated distortion results are significantly affected by failing to properly take into account the microstructural changes present in the deposited material [29]. Griffith et al. showed that the high temperatures reached during the laser deposition of stainless steel 316 can cause the material to anneal, thus reducing the measured residual stress [30]. Song et al. accounted for the plastic strain relaxation in a deposited nickel based alloy by setting a threshold temperature that, when surpassed, sets the plastic strain value to zero [31]. Qiao et al. performed micro-hardness measurements to demonstrate that the equivalent plastic strain needs to be dynamically adjusted at high temperatures in thermo-elasto-plastic models to simulate material annealing [32]. An approach for managing the stress relaxation in AM deposited Ti-6Al-4V has not yet been presented.

The objective of the present work is to develop a finite element model for predicting the *in situ* thermo-mechanical response of a Ti-6Al-4V workpiece deposited using a wire-fed electron beam system. Workpiece distortion and residual stress is modeled using a three-dimensional thermo-elasto-plastic analysis. The model is validated using experimental *in situ* temperature and distortion measurements performed during deposition of a 16 layer high, single bead wide, wall build as well as post-process residual stress measurements taken using Blind Hole Drilling. Both the *in situ* distortion and post-process residual stress measurements suggest that stress relaxation occurs in Ti-6Al-4V during deposition. The thermo-elasto-plastic model presented accounts for the observed stress relaxation by resetting both stress and plastic strain to zero when the temperature exceeds a prescribed stress relaxation temperature. The absorption efficiency, emissivity, and stress relaxation temperature are determined by applying inverse simulation.

9.2 ELECTRON BEAM DEPOSITION SIMULATION

The thermal and mechanical histories are determined by performing a three-dimensional transient thermal analysis and a three-dimensional quasi-static incremental analysis, respectively. The thermal and mechanical analyses are performed independently and are weakly coupled, meaning that the mechanical response has no effect on the thermal history of the workpiece [33].

9.2.1 Mechanical Analysis

A thermal history dependent quasi-static mechanical analysis is performed to obtain the mechanical response of the workpiece during deposition. The results of the thermal analysis are imported as a thermal load into the mechanical analysis.

For an incremental formulation, the constitutive law is re-written as:

$$^{n}\sigma = {}^{n-1}\sigma + \Delta\sigma \tag{9.1}$$

where $^{n-1}\sigma$ and $^{n}\sigma$ are the previous and current stress, and $\Delta\sigma$ is the stress increment computed as:

$$\Delta\sigma = \Delta\mathbf{C}(^{n-1}\boldsymbol{\varepsilon} - {}^{n-1}\boldsymbol{\varepsilon}_{\mathrm{p}} - {}^{n-1}\boldsymbol{\varepsilon}_{\mathrm{T}}) + \mathbf{C}(\Delta\boldsymbol{\varepsilon} - \Delta\boldsymbol{\varepsilon}_{\mathrm{p}} - \Delta\boldsymbol{\varepsilon}_{\mathrm{T}}) \tag{9.2}$$

where left superscripts denote the time increment over which a quantity is computed and $\Delta\boldsymbol{\varepsilon}$ is the total strain increment corresponding to the current displacement increment.

Incorporation of annealing in the constitutive system involves re-setting the plastic strain $\boldsymbol{\varepsilon}_{\mathrm{p}}$ to zero when the annealing temperature is reached [31]. Alternatively, Qiao et al. proposed a gradual (dynamic) reduction of the equivalent plastic strain $\boldsymbol{\varepsilon}_{\mathrm{q}}$ based on the time duration when the material is exposed at high temperatures [32]. As discussed in further detail in Sections 9.5.2 and 9.5.3 a relaxation mechanism is present during the deposition of Ti-6Al-4V manifested by the reduction of distortion recorded by *in situ* distortion measurements and reduced residual stress as measured by post-process Blind Hole Drilling. Implementation of such annealing models did not result into significant improvement in the correlation of computed results with *in situ* displacement and post-process residual stress measurements. The deviation was in the order of 500%.

A stress relaxation is proposed in this work where instantaneous annealing and creep occurs when a stress relaxation temperature T_{relax} is reached. The relaxation is implemented by re-setting the stress $^{n-1}\sigma$, elastic strain $^{n-1}\boldsymbol{\varepsilon}_{\mathrm{e}}$, thermal strain $^{n-1}\boldsymbol{\varepsilon}_{\mathrm{T}}$, plastic strain $^{n-1}\boldsymbol{\varepsilon}_{\mathrm{p}}$, and equivalent plastic strain $^{n-1}\boldsymbol{\varepsilon}_{\mathrm{q}}$ at the previous time increment to zero when the norm of the temperature at all Gauss points of an element exceeds T_{relax}. The material response corresponding to this relaxation model for various relaxation temperatures are presented in Section 9.5.

The Appendix displays the temperature dependent mechanical properties of Ti-6Al-4V, including the elastic modulus E, the yield strength σ_y, and the coefficient of thermal expansion α. Material properties between the temperatures listed are linearly interpolated. Mechanical properties are assumed constant above $800\,^{\circ}\mathrm{C}$. The Poison's ratio ν is assumed to be a constant value of 0.34. The model assumes perfect plasticity, i.e., no material hardening occurs.

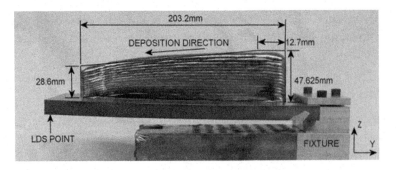

FIGURE 9.1 Sample mounted in the test fixture, and a schematic of the approximated deposited geometry.

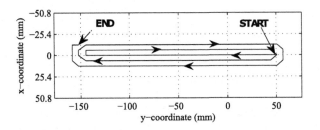

FIGURE 9.2 Scan pattern of the preheat performed on the top of the substrate.

9.3 CALIBRATION AND VALIDATION

9.3.1 Deposition Process

Electron beam freeform fabrication is used to deposit Ti-6Al-4V in a vacuum chamber. A Ti-6Al-4V plate 254 mm long, 101.6 mm wide, and 12.7 mm thick is used as a substrate. The substrate is clamped at one end and cantilevered, allowing the unconstrained end to deflect freely, while monitored by a laser displacement sensor (LDS). Figure 9.1 shows the test fixture and constrained substrate after deposition.

The AM system used is the Sciaky VX-300, which welds in the range of 10^{-4} to 10^{-5} Torr. The work envelope is approximately $5.8 \times 1.2 \times 1.2$ m^3 in volume. The deposition process begins by preheating the portion of the substrate where metal will be deposited. Figure 9.2 shows the preheating path. The electron beam power is set to 4.4 kW and scan speeds range between 35 mm/s and 42 mm/s.

After heating the top surface of the substrate a one-bead wide deposition is built 16 layers high and 203.2 mm long. The wire feed rate is set to 50.8 mm/s. The scan speed is a constant 12.7 mm/s. The electron beam operates at a power varying between 8 and 10 kW. The power is fluctuated to control the melt pool size. Figure 9.1 shows the resulting deposited geometry. The sloped wall is approximated as being 47.625 mm high near the clamped end of the substrate and 28.6 mm high near the free end for the succeeding finite element modeling work.

TABLE 9.1 Varying electron beam process and path parameters

Pass number	Power (kW)	Start time (s)	z start position (mm)	z end position (mm)
Preheat	4.4	0	0	0
1	8.8	115.72	2.9766	1.7875
2	8.7	216.02	5.9531	3.5750
3	9.3	297.03	8.9297	5.3625
4	8.7	436.69	11.9063	7.1500
5	8.6	535.33	14.8828	8.9375
6	8.6	624.49	17.8594	10.7250
7	8.6	736.94	20.8359	12.5125
8	8.4	833.99	23.8125	14.3000
9	8.3	941.90	26.7891	16.0875
10	8.4	1070.13	29.7656	17.8750
11	8.4	1195.32	32.7422	19.6625
12	8.3	1369.74	35.7188	21.4500
13	8.3	1524.81	38.6953	23.2375
14	8.4	1690.23	41.6719	25.0250
15	8.7	1990.64	44.6484	26.8125
16	8.4	2188.92	47.6250	28.6000

Table 9.1 shows, for each deposition layer, the average power and start time, as well as the beginning and ending z-coordinates.

9.3.2 *In Situ* Distortion and Temperature

In situ distortion measurements are taken using a Micro-Epsilon LDS, model LLT 28x0-100 and Micro-Epsilon controller scan CONTROL 28x0. The LDS is positioned to measure the longitudinal bowing distortion mode in the z-direction at the free end of the substrate, as shown in Figure 9.3B. The LDS targets a point approximately 6.3 mm from the free end of the substrate. A National Instruments 9250 module reads the LDS analog voltage signal.

Figure 9.3 shows the locations of the 4 thermocouples (0.25 mm diameter) used to monitor *in situ* temperature. Thermocouples 1–3 measure the temperature on the bottom of the substrate, parallel to the axis of deposition, and are located 63.5 mm from the clamped end of the substrate, at the center of the substrate, and 6.35 mm from the free end of the substrate, respectively. Thermocouple 4 measures temperature along the axis of deposition, on the top edge, at the free end of the substrate. The thermocouples are placed to allow for temperature readings at various substrate locations, without interfering with the LDS. A National Instruments 9213 module reads the thermocouple analog voltage signals. Both National Instruments modules record data into LabView at a frequency of 20 Hz.

9.3.3 Residual Stress

Post-process residual stress is measured using the hole-drilling method. Eight residual stress measurements are taken, seven measurements on the bottom of the substrate along the axis of deposition (see Figure 9.4A), and one measurement on the deposited material (see

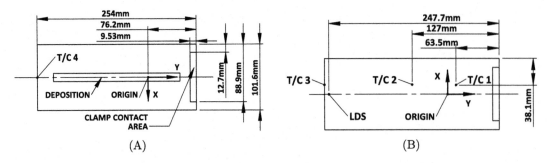

FIGURE 9.3 Schematic showing the LDS measurement location and the thermocouple (T/C) locations on the (A) top of the substrate and (B) the bottom of the substrate.

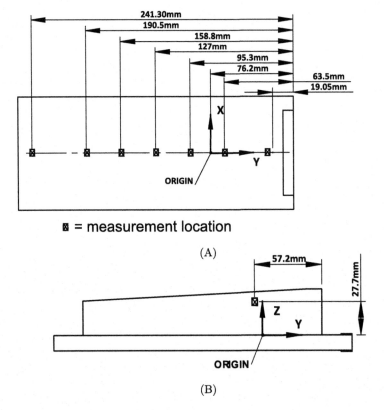

FIGURE 9.4 Schematic showing the locations of the residual stress measurements (A) on the bottom of the substrate and (B) on the build portion.

Figure 9.4B). The majority of the measurements are taken on the substrate, as it provides a large smooth surface appropriate for applying strain gauges and placing the milling guide. Micro-Measurements® strain gauges, model EA-06-062RE-120, are bonded to the bottom cen-

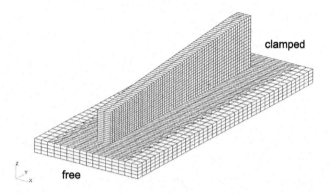

FIGURE 9.5 Mesh generated for the thermal and the mechanical analyses.

ter of the substrate. Gauges are calibrated using the procedure described in manufacturer engineering data sheet U059-07 and technical note 503. The ASTM E837 drilling process is followed. Incremental drilling is done using RS-200 Milling Guide and high speed drill from Micro-Measurements®. A 15.2 mm diameter, carbide-tipped, Type II Class 4A drill bit was used. Strain measurements are read by a Micro-Measurements® P-3500 Strain Indicator. Bridges are balanced with a Micro-Measurements® Switch and Balance Unit, model SB-1.

9.4 NUMERICAL IMPLEMENTATION

Figure 9.5 displays the 3-D finite element mesh used for both the thermal and mechanical analysis. The mesh contains 6848 Hex-8 elements and 9405 nodes. The mesh is generated using Patran 2012 by MSC. The mesh allots one element per deposition thickness and one element per heat source radius. A mesh convergence study was performed using two and then four elements per heat source radius resulting to 1.47% and 7.17% peak change respectively in the computed temperatures at the nodes corresponding to thermocouples and a 1.02% and 2.85% average change respectively at the nodes corresponding to thermocouples. However, the mesh with one element per heat source radius is used in this work because computational efficiency is critical in modeling electron beam deposition. The thermal and mechanical analyses are performed using Netfabb Simulation by Autodesk Inc. The quiet element activation approach is used in this work.

Figure 9.6 illustrates the mechanical constraints applied to the model. Three corner nodes on the clamped end of the substrate are constrained in all translational directions to model clamping the end of the substrate.

The numerical model is calibrated, using inverse simulation, as in reference [34], to determine the unknown values of efficiency η, surface emissivity ε, and stress relaxation temperature T_{relax}.

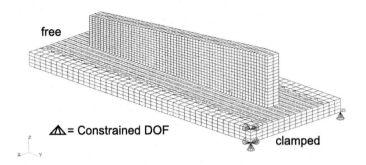

FIGURE 9.6 Mechanical constraints used to fix the clamped end of the substrate.

Efficiencies are varied from 0.90 [16] to 0.95 [35], based on the available literature. The emissivity is varied from 0.44 to 0.69 [36]. The percent error is calculated as:

$$\% \; Error = \frac{100 \sum_{i=1}^{n} | \frac{(x_{exp})_i - (x_{sim})_i}{(x_{exp})_i} |}{n} \tag{9.3}$$

where n is the total number of simulated time increments, i is the current time increment, x_{sim} is the simulated value, and x_{exp} is the experimentally measured value. The combination of efficiency and emissivity yielding the lowest percent error is used for the model.

The results from the calibrated thermal model are imported as a thermal load into the mechanical simulation. The mechanical model is calibrated by adjusting the relaxation temperature T_{relax}. Temperatures between 600 °C (below the beginning of the alpha-beta phase transition [37]) and 980 °C (approaching the beta-transus of Ti-6Al-4V) are tested. The error is calculated using Eq. (9.3).

9.5 RESULTS AND DISCUSSION

9.5.1 Thermal History

Table 9.2 shows the results of the simulated cases used to calibrate the thermal model. The values of efficiency η and emissivity ε are found to be 0.90 and 0.54, respectively, as this combination results in the lowest percent error (7.7%) compared with experiment. The calibrated value for efficiency agrees with that used for the electron beam deposition in reference [16]. The calibrated emissivity value falls within the range experimentally established by Coppa [36].

Figure 9.7 displays the thermal histories experimentally measured by the 4 thermocouples compared with the simulation results for the calibrated process efficiency of $\eta = 0.90$ and emissivity of $\varepsilon = 0.54$ at nodes corresponding to the locations of the thermocouples. A process time of 0 minutes corresponds to the start of the preheat. The ambient temperature is approximately 25 °C. Thermocouple 2 (located at the middle of the bottom of the substrate) records the highest temperature. Thermocouple 3 (located on the bottom of the substrate near

TABLE 9.2 Cases examined for thermal model calibration

Case	η	ε	% Error
1	0.90	0.44	12.1
2	0.90	0.49	9.2
3	0.90	0.54	7.7
4	0.90	0.59	8.1
5	0.90	0.64	9.5
6	0.90	0.69	11.1
7	0.925	0.44	13.6
8	0.925	0.49	10.5
9	0.925	0.54	8.8
10	0.925	0.59	8.9
11	0.925	0.64	9.9
12	0.925	0.69	10.5
13	0.95	0.44	15.3
14	0.95	0.49	12.1
15	0.95	0.54	10.1
16	0.95	0.59	9.8
17	0.95	0.64	10.3
18	0.95	0.69	11.5

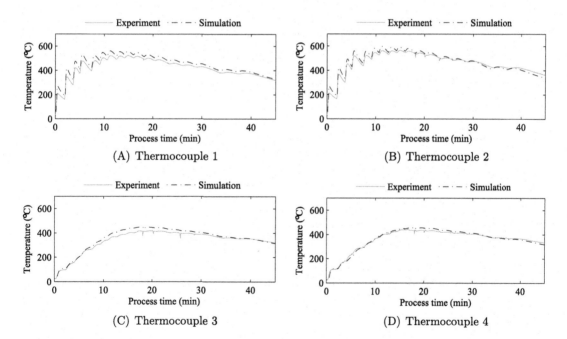

(A) Thermocouple 1

(B) Thermocouple 2

(C) Thermocouple 3

(D) Thermocouple 4

FIGURE 9.7 Computed thermal history with $\eta = 0.90$ $\varepsilon = 0.54$ compared with the experimental measurements.

TABLE 9.3 Cases examined for mechanical model calibration

Case	T_{relax} (°C)	% Error	Final % Error
19	600	43.4	45.3
20	650	16.9	23.2
21	670	6.6	12.9
22	680	4.7	8.0
23	690	7.4	2.2
24	700	14.3	4.4
25	980	144.3	157.5
26	None	431.0	520.3

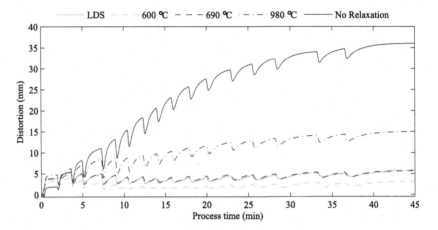

FIGURE 9.8 Computed distortion histories at various stress relaxation temperatures compared with the experimental measurement.

the clamped end) records the lowest temperature, as it is nearest to the clamp which absorbs heat through conduction. Although the process efficiency η and emissivity ε are calibrated to match the thermocouple measurements, a good correlation (7.7% error) is achieved for the entire duration of the process. In addition, thermocouples 1, 3, and 4, which are not used for calibration, also display close correlation with simulated results.

9.5.2 Distortion History

Table 9.3 lists the results of the simulated cases run to investigate the effect of changing the relaxation temperature on the *in situ* distortion. A final error is also computed by comparing the post-process experimental distortion with the post-process simulation distortion. A stress relaxation temperature of 690 °C provides results in closest agreement with the experimental post-process distortion.

Figure 9.8 illustrates the computed model distortion at the LDS point for selected relaxation temperatures. As seen in the figure, no relaxation (case 26) results in a nearly linear

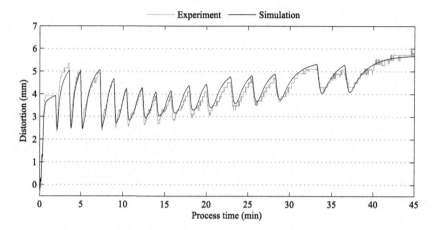

FIGURE 9.9 Computed *in situ* distortion at the LDS point compared with the experimental results.

increase of distortion with time up to 20 minutes. The increase rate becomes lower afterwards. The distortion trend is quite different from that of the LDS measurement which shows a decrease of distortion after 7 minutes of processing. Also, the final distortion is overpredicted by 520.3%. A high relaxation temperature (case 25, 980 °C) exhibits a similar distortion trend to that of with no relaxation, but with lower magnitudes. The distortion is still overpredicted by 157.5%. Case 23 with T_{relax} of 690 °C exhibits the same distortion behavior as the LDS measurement. The distortion decreases after 7 minutes of processing and starts to increase again at 20 minutes of processing time. The error compared to the LDS measurement is 7.4% over the entire process duration. An even lower relaxation temperature (case 19, 600 °C) exhibits a decrease of distortion after 5 minutes and an increase again after 25 minutes. This case underpredicts the final distortion, by 45.3%. These results illustrate the importance of accounting for stress relaxation in the model when predicting *in situ* and post-process distortion.

Figure 9.9 displays a close-up of the *in situ* measured distortion and the computed results for the calibrated relaxation temperature of 690 °C. It is noted that although the relaxation temperature is calibrated to match the *in situ* measured distortion, a very good correlation (7.4% error) is achieved for the entire duration of the process.

9.5.3 Residual Stress

Computed results for no stress relaxation (case 26) and stress relaxation temperature of $T_{relax} = 690$ °C (case 23) are shown in Figure 9.10. The results with no relaxation are overpredicted by more than 500%, and reach values as high as 1040 MPa. The computed residual stress also exhibits a shift to negative values close to the edges which is not present in the measured residual stress. The computed results with a stress relaxation temperature of $T_{relax} = 690$ °C do not show this shift to negative values and are in close agreement (within 25%) with the measured residual stress. The maximum stress predicted using stress relaxation is 41.3 MPa. The measured residual stress on the deposition is 20.0 MPa, which is in good agreement with the prediction from the model of 3.15 MPa considering that the mea-

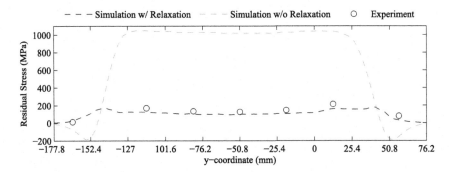

FIGURE 9.10 Computed residual stress results with and without stress relaxation, compared with the experimental measurements along the axis of deposition.

surement error of Blind Hole Drilling can be ±50 MPa [38]. The residual stress results on the deposition are lower than those on the substrate. This is because the deposited material reaches a temperature exceeding the stress relaxation temperature more frequently, leading to a reduction in residual stress. These results show that, in addition to causing errors in the simulated distortion, failure to account for stress relaxation present in Ti-6Al-4V during electron beam deposition results in erroneous residual stress predictions.

9.6 CONCLUSIONS

A finite element model is developed for predicting the thermo-mechanical response of Ti-6Al-4V during electron beam deposition. A three-dimensional thermo-elasto-plastic analysis is performed and experimental *in situ* temperature and distortion measurements are performed during deposition of a single bead wide, 16 layer high wall build for model validation. Post-process Blind Hole Drilling residual stress measurements are also performed.

Both the *in situ* distortion and post-process residual stress measurements suggest that stress relaxation occurs during the deposition of Ti-6Al-4V. A method of accounting for such stress relaxation in thermo-elasto-plastic simulations is proposed where both stress and plastic strain is reset to zero, when the temperature exceeds a prescribed stress relaxation temperature.

Inverse simulation is performed to determine the values of the absorption efficiency and the emissivity of electron beam deposited wire-fed Ti-6Al-4V, as well as the appropriate stress relaxation temperature. An efficiency of 0.90 and an emissivity of 0.54 result into the best correlation between measured and computed temperature history. Both values are in agreement to those published for laser assisted machining and powder based additive manufacturing. A stress relaxation temperature value of 690 °C is found to provide the best correlation between *in situ* measured and computed distortion. The results show that failure to implement stress relaxation in the constitutive model leads to errors in the residual stress and the *in situ* distortion predictions of over 500% when compared with the experimental measurements.

Suggested future work includes sectioning and microstructural analysis of the test piece and establishing a correlation with the computed results. In addition, the possibility of a gradual stress relaxation model should be investigated to compare with the instantaneous model used in this work. This would require further testing to establish the suitable rate of relaxation.

References

[1] Hibbitt HD, Marcal PV. A numerical, thermo-mechanical model for the welding and subsequent loading of a fabricated structure. Comput Struct 1973;3(5):1145–74.

[2] Friedman E. Thermomechanical analysis of the welding process using the finite element method. J Press Vessel Technol 1975;97:206.

[3] Argyris JH, Szimmat J, Willam KJ. Computational aspects of welding stress analysis. Comput Methods Appl Mech Eng 1982;33(1):635–65.

[4] Free J, Porter Goff R. Predicting residual stresses in multi-pass weldments with the finite element method. Comput Struct 1989;32(2):365–78.

[5] Tekriwal P, Mazumder J. Finite element analysis of three-dimensional transient heat transfer in GMA welding. Weld J 1988;67(5):150–6.

[6] Michaleris P, Tortorelli DA, Vidal CA. Analysis and optimization of weakly coupled thermoelastoplastic systems with applications to weldment design. Int J Numer Methods Eng 1995;38(8):1259–85.

[7] Lindgren LE, Runnemalm H, Näsström MO. Simulation of multipass welding of a thick plate. Int J Numer Methods Eng 1999;44(9):1301–16.

[8] Henwood C, Bibby M, Goldak J, et al. Coupled transient heat transfer-microstructure weld computations (Part B). Acta Metall 1988;36(11):3037–46.

[9] Leblond J, Devaux J, Devaux JC. Mathematical modelling of transformation plasticity in steels I: case of ideal-plastic phases. Int J Plast 1989;5(6):551–72.

[10] Lindgren LE. Finite element modeling and simulation of welding. Part 1: increased complexity. J Therm Stresses 2001;24(2):141–92.

[11] Lindgren LE. Finite element modeling and simulation of welding. Part 2: improved material modeling. J Therm Stresses 2001;24(3):195–231.

[12] Lindgren LE. Finite element modeling and simulation of welding. Part 3: efficiency and integration. J Therm Stresses 2001;24(4):305–34.

[13] Michaleris P, Zhang L, Bhide S, et al. Evaluation of 2D, 3D and applied plastic strain methods for predicting buckling welding distortion and residual stress. Sci Technol Weld Join 2006;11(6):707–16.

[14] Michaleris P. Modeling metal deposition in heat transfer analyses of additive manufacturing processes. Finite Elem Anal Des 2014;86:51–60.

[15] Yang J, Sun S, Brandt M, et al. Experimental investigation and 3D finite element prediction of the heat affected zone during laser assisted machining of Ti-6Al-4V alloy. J Mater Process Technol 2010;210(15):2215–22.

[16] Shen N, Chou K. Thermal modeling of electron beam additive manufacturing process–powder sintering effects. In: Proc. ASME 2012 int manuf sci and eng conf, 2012 June 4–8, Notre Dame, IN. ASME; 2012. p. 1–9.

[17] Dai K, Shaw L. Distortion minimization of laser-processed components through control of laser scanning patterns. Rapid Prototyping J 2002;8(5):270–6.

[18] Chiumenti M, Cervera M, Salmi A, et al. Finite element modeling of multi-pass welding and shaped metal deposition processes. Comput Methods Appl Mech Eng 2010;199(37):2343–59.

[19] Anca A, Fachinotti VD, Escobar-Palafox G, et al. Computational modelling of shaped metal deposition. Int J Numer Methods Eng 2011;85(1):84–106.

[20] Lundbäck A, Lindgren LE. Modelling of metal deposition. Finite Elem Anal Des 2011;47(10):1169–77.

[21] Picu R, Majorell A. Mechanical behavior of Ti-6Al-4V at high and moderate temperatures part II: constitutive modeling. Mater Sci Eng A 2002;326(2):306–16.

[22] Kelly S, Kampe S. Microstructural evolution in laser-deposited multilayer Ti-6Al-4V builds: part II. Thermal modeling. Metall Mater Trans A 2004;35(6):1869–79.

[23] Kelly S, Kampe S. Microstructural evolution in laser-deposited multilayer Ti-6Al-4V builds: part I. Microstructural characterization. Metall Mater Trans A 2004;35(6):1861–7.

[24] Qian L, Mei J, Liang J, et al. Influence of position and laser power on thermal history and microstructure of direct laser fabricated Ti-6Al-4V samples. Mater Sci Technol 2005;21(5):597–605.

[25] Fan Y, Cheng P, Yao Y, et al. Effect of phase transformations on laser forming of Ti-6Al-4V alloy. J Appl Phys 2005;98(1):013518.

[26] Peyre P, Aubry P, Fabbro R, et al. Analytical and numerical modelling of the direct metal deposition laser process. J Phys D, Appl Phys 2008;41(2):025403.

[27] Ahsan MN, Pinkerton AJ. An analytical–numerical model of laser direct metal deposition track and microstructure formation. Model Simul Mater Sci Eng 2011;19(5):055003.

[28] Longuet A, Robert Aeby-Gautier Y, et al. A multiphase mechanical model for Ti-6Al-4V: application to the modeling of laser assisted processing. Comput Mater Sci 2009;46(3):761–6.

[29] Ghosh S, Choi J. Three-dimensional transient finite element analysis for residual stresses in the laser aided direct metal/material deposition process. J Laser Appl 2005;17:144.

[30] Griffith M, Schlienger M, Harwell L. Thermal behavior in the LENS process. In: Proceedings of the solid freeform fabrication symposium. University of Texas at Austing Publishers; 1998. p. 89–97.

[31] Song X, Xie M, Hofmann F, et al. Residual stresses and microstructure in powder bed direct laser deposition (PB DLD) samples. Int J Mater Form 2014;10:1–10.

[32] Qiao D, Feng Z, Zhang W, et al. Modeling of weld residual plastic strain and stress in dissimilar metal butt weld in nuclear reactors. In: Proc. ASME 2013 press vess & piping div conf. ASME; 2013. PVP2013–98081.

[33] Zhang L, Reutzel E, Michaleris P. Finite element modeling discretization requirements for the laser forming process. Int J Mech Sci 2004;46(4):623–37.

[34] Aarbogh HM, Hamide M, Fjaer HG, et al. Experimental validation of finite element codes for welding deformations. J Mater Process Technol 2010;210(13):1681–9.

[35] Taminger KM, Hafley RA. Electron beam freeform fabrication: a rapid metal deposition process. In: Proc 3rd ann automotive compos conf. Society of Plastics Engineers; 2003. p. 9–10.

[36] Coppa P, Consorti A. Normal emissivity of samples surrounded by surfaces at diverse temperatures. Measurement 2005;38(2):124–31.

[37] Elmer J, Palmer T, Babu S, et al. In situ observations of lattice expansion and transformation rates of α and β phases in Ti-6Al-4V. Mater Sci Eng A 2005;391(1):104–13.

[38] Withers P, Bhadeshia H. Residual stress part 1–measurement techniques. Mater Sci Technol 2001;17(4):355–65.

Thermo-Mechanical Modeling of Large Electron Beam Builds*

Erik R. Denlinger

Product Development Group, Autodesk Inc., State College, PA, United States

10.1 INTRODUCTION

Electron beam directed energy deposition is a commonly used additive manufacturing (AM) process. The process is of particular interest to the aerospace industry, due to its ability to rapidly deposit bulk material at rates as high as 2500 cm^3/hr [1]. The electron beam process constructs the part on a layer by layer basis, directly from a digital drawing file. Wire feedstock material is rapidly heated until melting and then allowed to solidify and cool, forming a fully dense geometry. The repeated thermal cycling causes large thermal gradients which, in turn, result in undesirable levels of distortion in the workpiece. Targeted alterations to the build plan may be used to reduce distortion levels. To avoid expensive trial and error iterations, an accurate predictive model is needed to optimize the build plan. Electron beam deposited parts are typically large, involving the deposition of several hundred kilograms of material, resulting in computationally expensive finite element models. Thus, for large parts, a modeling strategy is required to reduce computation time in order to make the simulations feasible.

Thermal and mechanical modeling of AM is similar to that of multi-pass welding, which has been an active research area for nearly 40 years. Numerous weld models capable of predicting thermal and mechanical behavior are reported in references [2–11]. State of the art summaries by Lindgren et al. detail the development of weld model complexity [12], weld material modeling [13], and weld model computational efficiency [14].

Modeling of AM processes adds significant computational expense when compared to multi-pass weld simulations. This is due to an increase in the process time and the number of passes, as well as the addition of the deposited material. The deposition of material can add a large number of elements into the simulation and requires an element activation strategy [15].

---

* This chapter is based upon the original work: Denlinger, Erik R., Jeff Irwin, and Pan Michaleris. "Thermomechanical modeling of additive manufacturing large parts." Journal of Manufacturing Science and Engineering 136.6 (2014): 061007.

Modeling of AM is primarily focused on predicting thermal response [16–21], predicting distortion and residual stress, [22–26], and developing distortion mitigation techniques [27,28].

All aforementioned models focus on the simulation of small parts with a simple deposited geometry to insure feasibility of computation. The capability to model large parts is of use in industry applications where parts are commonly large and geometries complex. The use of 2D models is one approach used to minimize computational expense. This strategy can result in an increasing loss of accuracy with added weld passes and would not be suitable for complex geometries [29]. Another approach is to perform a steady-state analysis rather than a transient analysis. In the area of welding research, Zhang and Michaleris developed a finite element model using an Eulerian approach which decreased computation time by a factor of 2. The approach, however, resulted in a loss of accuracy when compared with a Lagrangian approach [30]. For AM processes, Ding et al. presents an Eulerian simulation capable of simulating the deposition of four 0.5 m long beads in under 20% of the computation time required to do the same simulation using a transient approach [31]. The use of an Eulerian reference frame allows for a nonuniform mesh along the deposition line, resulting in reduced computation time, but prohibits the modeling of more complex geometries.

Adaptive meshing can also be used to reduce computation time [32,33]. Work has been done on adaptive meshing outside the field of additive manufacturing, especially in computational fluid dynamics where singularities need to be resolved. Most techniques first use an error estimator [34] or error indicator [35] to determine which regions of the mesh need to be refined or coarsened. Then, the necessary changes are made to the mesh and the solution is transferred to new grid points. Many different techniques exist. Berger and Oliger used adaptive subgrid generation to refine meshes for finite difference solutions of 1- and 2-dimensional hyperbolic partial differential equations [32]. They showed that adaptive meshing can yield more accurate results than fine static meshing with shorter computation times. Berger and Colella further developed this method in [36]. Bell et al. extended this work to 3 dimensions [37]. Bank et al. developed algorithms for refining quadrilateral meshes which can be generalized to hexahedral meshes [38]. Shepherd et al. developed algorithms for coarsening structured and unstructured hexahedral meshes [39].

In the field of welding, Prasad and Narayanan developed a triangular adaptive mesh method for 2-dimensional thermal models [40]. Runnemalm and Hyun presented a fully automatic 3-dimensional thermo-mechanical adaptive technique for weld modeling with hexahedral elements [41], but were not able to assess the accuracy or the efficiency of their technique due to the high physical memory demands of a benchmark fine mesh. Unfortunately there is currently no model presented in the available literature capable of simulating the deposition of many layers of complex geometry for the manufacture of a large (on the order of meters) workpiece.

Previous work performed by the authors presents a method for calibrating and validating a model for simulating the thermal and mechanical response of an electron beam deposited titanium workpiece [42]. A single bead wide 16 layer deposition was manufactured while monitoring substrate temperatures using in-situ thermocouple measurements and in-situ substrate distortion using a laser displacement sensor (LDS). A finite element model was used to identify the absorption efficiency and the emissivity, as well as the stress relaxation temperature for Ti-6Al-4V. The model was further validated by comparing measured and predicted residual stress values along the substrate. The values for absorption efficiency, emissivity, and stress relaxation temperature found in reference [42] are applied in this work with no further calibration.

The objective of this work is to develop a finite element modeling strategy to allow for the prediction of the distortion accumulation in large AM parts. A thermo-mechanical finite element analysis is performed using a hybrid quiet inactive element activation strategy combined with adaptive coarsening. At the beginning of the simulation, before material deposition commences, elements corresponding to deposition material are removed from the analysis, then elements are introduced in the model layer by layer in a quiet state with material properties rendering them irrelevant. As the moving energy source is applied on the part, elements are switched to active by restoring the actual material properties when the energy source is applied on them. A layer by layer coarsening strategy merging elements in lower layers of the build is also implemented such that while elements are added on the top, elements are also merged on the bottom maintaining a low number of degrees of freedom in model for the entire simulation. The algorithm requires less overhead than a fully automatic scheme, but requires no manual mesh modification as in reference [43]. The effectiveness of the approach is demonstrated by modeling the manufacturing of a large Ti-6Al-4V part consisting of 107 total layers and several thousand weld passes. Model validation is performed using experimental measurements. The large part is built by Sciaky Inc., and the distortion measurements are taken by Neomek Inc.

10.2 ELECTRON BEAM DEPOSITION SIMULATION

As demonstrated in Chapter 2, the thermal and mechanical histories are determined by performing a 3D transient thermal analysis and a 3D quasi-static incremental analysis, respectively. The thermal analysis is performed independently of the mechanical analysis, as the mechanical response has no effect on the thermal history of the workpiece [44].

10.3 MESH COARSENING ALGORITHM

10.3.1 Merging of Elements Layer by Layer

To reduce computational expense, a coarsening algorithm is developed for large models. Layers are merged two layers below active deposition, as shown in Figure 10.1, such that there is always at least one fine layer below the deposition. The coarsening algorithm is combined with a hybrid quiet inactive element activation method that adds elements for the new layer. Nodes are removed and DOFs are deleted in the merged layers. For mechanical analysis, Gauss point variables such as plastic strain and hardening need to be interpolated.

10.3.2 Interpolation of Gauss Point Values

In order to interpolate the solution values to the Gauss points of the new element, the global coordinates of the Gauss points must be transformed to the local coordinates of each

FIGURE 10.1 The mesh coarsening algorithm merges two layers of elements and deletes an entire plane of nodes. This is combined with a hybrid quiet inactive element activation method.

of the two old elements. Transforming local coordinates to global coordinates, known as isoparametric mapping, can be accomplished using the shape functions, as shown in Equation (10.1) for linear hexahedral elements [45]:

$$\phi = \frac{1}{8} \sum_{i=1}^{8} \left[\Phi_i \prod_{k=1}^{3} (1 + \xi_k \Xi_{ki}) \right] \tag{10.1}$$

where ϕ is the variable being interpolated, $\boldsymbol{\Phi}$ are the values of ϕ at the element's nodes, $\boldsymbol{\xi}$ are the local coordinates of the point where the quantity is being interpolated, and $\boldsymbol{\Xi}$ are the local coordinates of the element's nodes. Subscript i ranges from 1 to 8 for each node of the element, and k ranges from 1 to 3 for each spatial dimension.

For prismatic elements, a simple linear interpolation could be used to transform global coordinates to local coordinates. However, for a general element, a more sophisticated method is needed. In this case, the Newton–Raphson method is used. This is an inverse isoparametric mapping, first described by Lee and Bathe for 2-dimensional elements [46]. Here, the procedure is described in detail for 3-dimensional hexahedrons. By substituting global coordinates x for ϕ and global node coordinates X for $\boldsymbol{\Phi}$ into Equation (10.1), the residual $\boldsymbol{R}$ can be formulated for $j = 1$ to 3 as follows:

$$R_j = x_j - \frac{1}{8} \sum_{i=1}^{8} \left[X_{ij} \prod_{k=1}^{3} (1 + \xi_k \Xi_{ki}) \right] \tag{10.2}$$

Then the Jacobian $\boldsymbol{J}$ can be formulated as shown in Equation (10.3):

$$J_{mj} = \frac{\partial R_j}{\partial \xi_m} = -\frac{1}{8} \sum_{i=1}^{8} \frac{X_{ij} \, \Xi_{mi} \prod_{k=1}^{3} (1 + \xi_k \Xi_{ki})}{1 + \xi_m \Xi_{mi}} \tag{10.3}$$

Beginning with $\boldsymbol{\xi}^0 = \boldsymbol{0}$, the Newton–Raphson scheme is iteratively applied as follows:

$$\boldsymbol{\xi}^{n+1} = \boldsymbol{\xi}^n - \left[\boldsymbol{J}^n \right]^{-1} \boldsymbol{R}^n \tag{10.4}$$

updating the residual and the Jacobian at each step for $n = 0, 1, 2, \ldots$ until the residual is sufficiently small. Note that $\boldsymbol{J}$ is only a 3×3 matrix, thus solving Equation (10.4) is computationally inexpensive. For the models tested here, this method always converges.

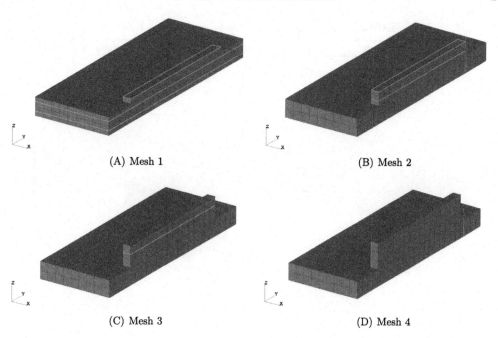

(A) Mesh 1   (B) Mesh 2

(C) Mesh 3   (D) Mesh 4

FIGURE 10.2 Mesh 1 has a uniform density in the z direction. Element edges which are subsequently eliminated by coarsening are highlighted in light grey.

Because the solution values are known at the Gauss points of the new elements rather than at the nodes, a modified version of Equation (10.1) must be used:

$$\phi = \frac{1}{8} \sum_{i=1}^{8} \left[\gamma_i \prod_{k=1}^{3} \left(1 + \sqrt{3}\, \xi_k \, \Xi_{ki} \right) \right] \tag{10.5}$$

where γ contains the values of ϕ at the Gauss points and the $\sqrt{3}$ accounts for the fact that the Gauss points for linear hexahedral elements are located at local coordinates of $\pm\frac{1}{\sqrt{3}}$. By substituting the local coordinates ξ obtained from the Newton–Raphson method into Equation (10.5), the six stress components and other solution values can be interpolated to the new elements.

10.3.3 Verification of Layer by Layer Coarsening Algorithm

A small scale model is used to verify the coarsening algorithm. Thermo-mechanical modeling of electron beam deposition is performed on the small model using mesh density and processing conditions similar to that of modeling a large part. Results from a model without coarsening are used as a baseline. The four successively coarser meshes used in each stage of the simulation are shown in Figure 10.2. The numbers of nodes and elements in each mesh, including the full wall height, are shown in Table 10.1.

TABLE 10.1   Number of nodes and elements in each mesh

Mesh	Number of nodes	Number of elements
1	2261	1632
2	1479	912
3	1428	880
4	1377	848

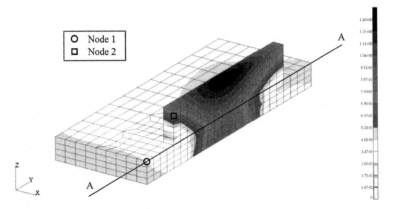

FIGURE 10.3   Displacement magnitude results (mm) at the end of simulation for the uniformly fine baseline case of the small model.

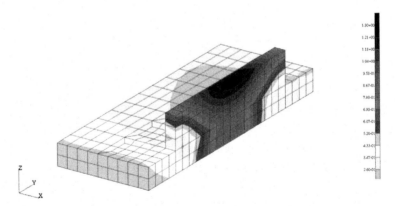

FIGURE 10.4   Displacement magnitude results (mm) at the end of simulation for the coarse case of the small model.

10.3.4 Verification Results

Displacement magnitude results at the end of the simulation are shown in Figure 10.3 and Figure 10.4 respectively for the uniform fine baseline mesh and the coarse mesh. Displacements at node 1 at the free end of the substrate and node 2 between the fifth and sixth layers are plotted versus time in Figure 10.5 for both cases. Note that node 1 is active for

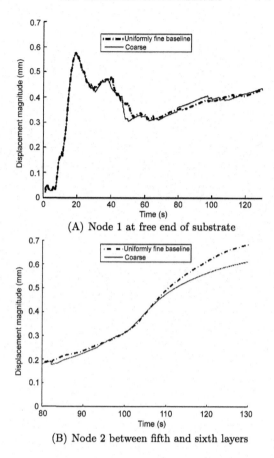

(A) Node 1 at free end of substrate

(B) Node 2 between fifth and sixth layers

FIGURE 10.5 Displacement magnitude at two different nodes versus time for both cases of the small model. The locations of these nodes are shown in Figure 10.3.

the duration of the simulation, but node 2 does not become active until approximately 82 s. Displacement results at the instant shown in Figure 10.3 along line AA are extracted and plotted in Figure 10.6 for both cases. The largest percent error for the results shown is 24.4% at $y = 44.45$ mm in Figure 10.6. This location corresponds to the end of the deposition track meaning that bending of the substrate ends at this point. The finer mesh is more able to capture this transition. The simulation wall times for both cases are compared in Table 10.2. There is no improvement in the first stage because at this point both meshes are the same.

10.4 VALIDATION ON A LARGE PART

Electron beam freeform fabrication is used to deposit Ti-6Al-4V wire feedstock material in a vacuum chamber, to form the workpiece shown in Figure 10.7. A Ti-6Al-4V plate 3810 mm

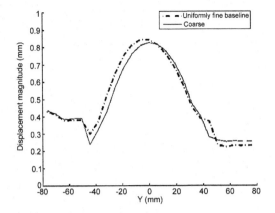

FIGURE 10.6 Displacement versus *y* location at end of simulation along line AA for both cases of the small model.

FIGURE 10.7 Large workpiece, deposited on 3810 mm long substrate, for model validation, figure provided by Sciaky, Inc.

TABLE 10.2 Quasistatic mechanical simulation wall times for the small model

Simulation stage	Uniformly fine baseline mesh	Coarse mesh	Percent reduction
1	17.4 s	17.4 s	0%
2	23.2 s	15.2 s	34.7%
3	23.5 s	9.69 s	58.8%
4	23.9 s	8.72 s	63.5%

long, 457 mm wide, and 25.4 mm thick is used as a substrate. The substrate is placed on the fixture and held in place by 40 evenly spaced clamps. The clamps have a spring constant of 22.5 N/mm, allowing for some distortion of the substrate.

The AM system used is the Sciaky VX-300, which welds in the range of 10^{-4} to 10^{-5} Torr. The work envelope is approximately $5.8 \times 1.2 \times 1.2 \ m^3$ in volume. The electron beam power is varied from 8 kW to 10 kW in order to control the melt pool size. The scan speed of the electron beam is 12.7 mm/s. The 12.7 mm diameter wire has a feed rate set to 50.8 mm/s. Previous work by the authors found the absorption efficiency η to be 0.90 [42]. The largest builds are deposited 80 layers high, with a total deposition layer count of 107. A post-process 3D scan of the part is performed by Neomek Inc. using a Surphaser Laser with and estimated scanning accuracy of ± 0.5 mm, which allows for the distortion levels to be quantified.

10.5 NUMERICAL IMPLEMENTATION

Due to the size and complexity of the large part, a 3-stage modeling approach is implemented to simulate the entire deposition process. Stage 1 models the first layer of deposition, stage 2 models the second through ninth layers of deposition, and stage 3 models all succeeding layers. This approach allows for mesh coarsening to be implemented between each stage, thus reducing the computational expense. All meshes are generated using Patran 2012 by MSC. Anti-symmetry is also used to model only half of the deposition. The thermal and mechanical analysis are performed using Netfabb Simulation by Autodesk Inc. The elements are activated using a hybrid quiet inactive approach for all simulations [15]. The inactive elements are initially not part of the simulation, but rather, are added on a layer by layer basis requiring the addition of an equation to the linear system in conjunction with a renumbering of the equations. The initially inactive elements are brought into the simulation as quiet elements, meaning that they are given material properties such that they do not affect the thermal or mechanical model. When the heat source contacts a quiet element it is activated by switching the element's material properties to their actual value.

The electron beam heat source is modeled using the Goldak double ellipsoid model [47] as follows:

$$Q = \frac{6\sqrt{3}P\eta f}{abc\pi\sqrt{\pi}}\, e^{-[\frac{3x^2}{a^2} + \frac{3y^2}{b^2} + \frac{3(z+v_w t)^2}{c^2}]} \tag{10.6}$$

where P is the power, η is the absorption efficiency, f is the process scaling factor; x, y, and z are the local coordinates; a, b, and c are the transverse, melt pool depth, and longitudinal dimensions of the ellipsoid respectively, v_w is the heat source travel speed, and t is the time. The front quadrant and rear quadrant of the heat source are modeled separately. The heat source radius a is 6.35 mm, and the melt pool depth b is 3.8 mm, for each quadrant. The length c measures 12.7 mm and 25.4 mm for the front and rear quadrant, respectively. The process scaling factor f equals 0.6 for the front quadrant and 1.4 for the rear quadrant. The elongation of the heat source allows for larger time steps and thus, reduced computation time. Velocity is a constant 12.7 mm/s. Power is entered as the power used during deposition, which varies between 8 and 10 kW. The absorption efficiency η is set at 0.90. Details on the determination of these parameters are discussed in Ref. [42].

All free surfaces are subject to radiative heat loss, including those located at the interface of active and quiet elements. Figure 10.8A shows the mechanical constraints applied to the model. Rigid constraints are placed to prevent the symmetry face from moving in the x direction. A constrained degree of freedom is placed on a single node at the symmetry face to prevent rigid-body motion in the y direction. Spring constraints are used to simulate the clamps of the fixture. The spring constant is set to a large value for negative deflections, simulating the rigid fixture, and is set to 11.25 N/mm for positive deflections, simulating the clamps present on the fixture. Table 10.3 lists the nonlinear spring constant values. Figure 10.8B shows a magnified isometric view of a portion of the mesh.

The first modeling stage models the preheat (i.e. initial heating passes performed before any deposition to make the substrate surface uniform) and the first layer of deposition on each side of the substrate. Stage 1 accounts for 79.3 hours of experimental process time, rep-

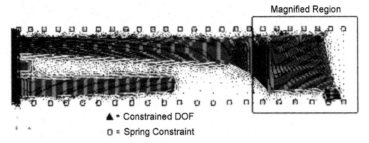

▲ - Constrained DOF

□ = Spring Constraint

(A) Top view of mesh with mechanical constraints

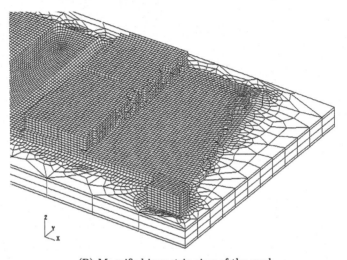

(B) Magnified isometric view of the mesh

FIGURE 10.8 Illustration of the mesh used for layers 1–9.

TABLE 10.3 Spring constraint properties of the fixture

z-deflection (mm)	Spring Constant (N/mm)
−6	500
−4	170
−2	100
>=0	11.25

resenting periods of material deposition and cooling. Mesh coarsening cannot be performed before the deposition of layer 1 is completed.

The substrate for stage 1 is 4 elements thick, with the top and bottom layer having a thickness of 3.175 mm (equal to the thickness of each deposition layer), and the middle layers having a thickness of 9.925 mm. The mesh for the deposited layers measures 3.175 mm thick, allowing for 1 element per deposition thickness. The mesh is not uniform. Deposited elements are dimensioned to provide 1 element per heat source radius [42], resulting in an element size

TABLE 10.4 Number of nodes, number of elements, and process time for each model

Stage	Number of nodes	Number of elements	CPU wall time (therm/mech (hrs.))
1	179,939	154,104	13.8/25.33
2	145,883	120,174	16.9/32.2
3	120,941	104,025	12.7/13.2

of roughly 6.35 mm × 6.35 mm. The mesh is coarsened outward from the deposition in order to reduce the number of required elements. The model contains 154,104 hexagonal elements and 179,939 nodes.

The second modeling stage models layers 2 through 9 of the deposition, and simulates 121.0 hours of process time. Mesh coarsening can be utilized after the completion of layer 1 and before the start of the deposition of layer 2. Once the deposition of layer 9 is completed, the majority of the material comprising the part has been deposited.

The substrate elements are coarsened in the z-direction for stage 2. This results in a 50% reduction in the number of elements located in the substrate. The model contains 120,174 hexagonal elements and 145,883 nodes. Note that due to the implementation of the mesh coarsening algorithm, the number of required elements for this model has decreased when compared with its predecessor, despite the increase in the amount of the deposited material.

The third modeling stage begins after the completion of the deposition of layer 9. The majority of the material has now been deposited. The remaining deposition completes the tallest structures present on the part, and accounts for 431.1 hours of process time. The deposition passes become shorter, and the heating caused by the electron beam becomes more localized.

Stage 3 simulates layers 10 through completion of the deposition. All elements that are not part of the tall structures have been coarsened. The mesh is now comprised of 104,025 hexagonal elements and 120,941 nodes. The model is run independently of the results of the previous 2 simulations in order to examine if the additional deposition layers have a significant impact on the overall substrate distortion. Some small structures have been omitted from the model as they will have a negligible effect on distortion compared with the larger structures.

10.6 RESULTS AND DISCUSSION

Table 10.4 displays the CPU run time for each of the 3 models run using 16, 3.1 GHz, cores. The thermal models for stages 1, 2, and 3 run in 13.8, 16.9, and 12.7 hours, respectively. The mechanical models run in 25.33, 32.2, and 13.2 hours, respectively. Stage 2 is the most computationally expensive because, while the model does contain fewer elements than that of stage 1, stage 2 simulates significantly more process time.

Figure 10.9 shows the final simulation results for the large part, which are compared to the 3D scan distortion results taken by Neomek Inc. Figure 10.10 shows the coordinate system that is used to compare the 3D scan distortion results to the computed distortion results, as well as the location of the data points being compared. The distortion is compared along the x direction due to the fact that it is the coordinate direction along which distortion varies most greatly. The face with the maximum distortion was chosen for comparison.

FIGURE 10.9 Displacement magnitude (mm) results after the model has been rotated to the same orientation as the scan results (2 × magnification).

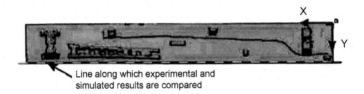

FIGURE 10.10 Top view of the coordinate system used for the simulated and experimental results, figure provided by Neomek, Inc.

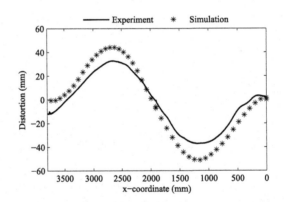

FIGURE 10.11 Experimental and simulated distortion results in the x–z plane at $y = 457$ mm.

Results of the simulation of the distortion accumulated from layer 10 to the completion of the deposition show that distortion attributed to these layers is 0.025 mm. These layers cause negligible distortion of the substrate, showing that nearly all substrate distortion is attributable to the first 9 layers of deposition. This result is expected, as a majority of the feedstock material is deposited during the first 9 layers of the deposition. The succeeding layers are deposited using shorter deposition passes, resulting in more localized thermal gradients.

Figure 10.11 displays a comparison between the simulated results and the actual experimental results. The experimental scan results and simulation results plotted correspond to distortions in the x–z plane at $y = 457$ mm. The simulated and experimental results show good agreement, with a maximum error of 29%.

10.7 CONCLUSIONS

A finite element modeling strategy has been developed to allow for the prediction of distortion accumulation in large workpieces in additive manufacturing. The strategy involves performing a 3D Lagrangian thermo-elasto-plastic analysis using a combined hybrid quiet inactive element activation strategy with adaptive coarsening. The effectiveness of the modeling strategy is demonstrated and experimentally validated on a large electron beam deposited Ti-6Al-4V part consisting of 107 deposition layers.

The proposed modeling strategy is particularly suited for models with a high number of elements, as the computational overhead to introduce new elements and to coarsen existing elements is negligible compared to the computational savings of the reduced total number of degrees of freedom in the model. Unlike an Eulerian approach the presented modeling strategy is capable of modeling complex geometries which are likely to appear in industry applications.

Model validation using experimental measurements shows that the proposed strategy can accurately predict the distortion of additive manufacturing large parts (maximum error of 29%). It can also enable determining which deposition passes are responsible for the majority of the substrate distortion what is important in designing build plans that mitigate distortion.

Possible future work could include implementing an octree-based element refining/coarsening strategy for structured or general unstructured hexahedral meshes. This could include an automatic algorithm that refines in the vicinity of the electron beam or laser path, or an error estimator/indicator method. Such an approach could significantly further reduce required computation time.

References

[1] Taminger KM, Hafley RA. Electron beam freeform fabrication: a rapid metal deposition process. In: Proc. 3rd ann automotive compos conf; 2003 Sept 9–10. Troy, MI: Society of Plastics Engineers; 2003. p. 9–10.

[2] Hibbitt HD, Marcal PV. A numerical, thermo-mechanical model for the welding and subsequent loading of a fabricated structure. Comput Struct 1973;3(5):1145–74.

[3] Friedman E. Thermomechanical analysis of the welding process using the finite element method. J Press Vessel Technol 1975;97:206.

[4] Andersson B. Thermal stresses in a submerged-arc welded joint considering phase transformations. J Eng Mater Technol (United States) 1978;100.

[5] Argyris JH, Szimmat J, Willam KJ. Computational aspects of welding stress analysis. Comput Methods Appl Mech Eng 1982;33(1):635–65.

[6] Papazoglou V, Masubuchi K. Numerical analysis of thermal stresses during welding including phase transformation effects. J Press Vessel Technol (United States) 1982;104(3).

[7] Free J, Porter Goff R. Predicting residual stresses in multi-pass weldments with the finite element method. Comput Struct 1989;32(2):365–78.

 [8] Tekriwal P, Mazumder J. Finite element analysis of three-dimensional transient heat transfer in GMA welding. Weld J 1988;67(5):150–6.
 [9] Michaleris P, Tortorelli DA, Vidal CA. Analysis and optimization of weakly coupled thermoelastoplastic systems with applications to weldment design. Int J Numer Methods Eng 1995;38(8):1259–85.
[10] Lindgren LE, Runnemalm H, Näsström MO. Simulation of multipass welding of a thick plate. Int J Numer Methods Eng 1999;44(9):1301–16.
[11] Asadi M, Goldak JA. An integrated computational welding mechanics with direct-search optimization for mitigation of distortion in an aluminum bar using side heating. J Manuf Sci Eng 2014;136(1):011007.
[12] Lindgren LE. Finite element modeling and simulation of welding. Part 1: increased complexity. J Therm Stresses 2001;24(2):141–92.
[13] Lindgren LE. Finite element modeling and simulation of welding. Part 2: improved material modeling. J Therm Stresses 2001;24(3):195–231.
[14] Lindgren LE. Finite element modeling and simulation of welding. Part 3: efficiency and integration. J Therm Stresses 2001;24(4):305–34.
[15] Michaleris P. Modeling metal deposition in heat transfer analyses of additive manufacturing processes. Finite Elem Anal Des 2014;86:51–60.
[16] Kolossov S, Boillat E, Glardon R, Fischer P, Locher M. 3D FE simulation for temperature evolution in the selective laser sintering process. Int J Mach Tools Manuf 2004;44(2):117–23.
[17] Peyre P, Aubry P, Fabbro R, et al. Analytical and numerical modelling of the direct metal deposition laser process. J Phys D, Appl Phys 2008;41(2):025403.
[18] Qian L, Mei J, Liang J, et al. Influence of position and laser power on thermal history and microstructure of direct laser fabricated Ti-6Al-4V samples. Mater Sci Technol 2005;21(5):597–605.
[19] Shen N, Chou K. Thermal modeling of electron beam additive manufacturing process–powder sintering effects. In: Proc. ASME 2012 int manuf sci and eng conf, 2012 June 4–8. Notre Dame, IN: ASME; 2012. p. 1–9.
[20] Jamshidinia M, Kong F, Kovacevic R. Numerical modeling of heat distribution in the electron beam melting® of Ti-6Al-4V. J Manuf Sci Eng 2013;135(6):061010.
[21] Sammons PM, Bristow DA, Landers RG. Height dependent laser metal deposition process modeling. J Manuf Sci Eng 2013;135(5):054501.
[22] Anca A, Fachinotti VD, Escobar-Palafox G, et al. Computational modelling of shaped metal deposition. Int J Numer Methods Eng 2011;85(1):84–106.
[23] Chiumenti M, Cervera M, Salmi A, et al. Finite element modeling of multi-pass welding and shaped metal deposition processes. Comput Methods Appl Mech Eng 2010;199(37):2343–59.
[24] Lundbäck A, Lindgren LE. Modelling of metal deposition. Finite Elem Anal Des 2011;47(10):1169–77.
[25] Marimuthu S, Clark D, Allen J, Kamara A, Mativenga P, Li L, et al. Finite element modelling of substrate thermal distortion in direct laser additive manufacture of an aero-engine component. J Mech Eng Sci 2012.
[26] Mughal M, Fawad H, Mufti R. Three-dimensional finite-element modelling of deformation in weld-based rapid prototyping. J Mech Eng Sci 2006;220(6):875–85.
[27] Chin R, Beuth J, Amon C. Control of residual thermal stresses in shape deposition manufacturing. In: Proceedings of the solid freeform fabrication symposium; 1995. p. 221–8.
[28] Klingbeil N, Beuth J, Chin R, Amon C. Residual stress-induced warping in direct metal solid freeform fabrication. Int J Mech Sci 2002;44(1):57–77.
[29] Michaleris P, Feng Z, Campbell G. Evaluation of 2D and 3D FEA models for predicting residual stress and distortion. ASME-PUBLICATIONS-PVP 1997;347:91–102.
[30] Zhang L, Michaleris P. Investigation of Lagrangian and Eulerian finite element methods for modeling the laser forming process. Finite Elem Anal Des 2004;40(4):383–405.
[31] Ding J, Colegrove P, Mehnen J, Ganguly S, Sequeira Almeida P, Wang F, et al. Thermo-mechanical analysis of wire and arc additive layer manufacturing process on large multi-layer parts. Comput Mater Sci 2011;50(12):3315–22.
[32] Berger MJ, Oliger J. Adaptive mesh refinement for hyperbolic partial differential equations. J Comput Phys 1984;53(3):484–512.
[33] Jasak H, Gosman A. Automatic resolution control for the finite-volume method, part 1: a-posteriori error estimates. Numer Heat Transf, Part B, Fundam 2000;38(3):237–56.
[34] Zienkiewicz OC, Zhu JZ. A simple error estimator and adaptive procedure for practical engineering analysis. Int J Numer Methods Eng 1987;24(2):337–57.

[35] Picasso M. An anisotropic error indicator based on Zienkiewicz–Zhu error estimator: application to elliptic and parabolic problems. SIAM J Sci Comput 2003;24(4):1328–55.

[36] Berger MJ, Colella P. Local adaptive mesh refinement for shock hydrodynamics. J Comput Phys 1989;82(1):64–84.

[37] Bell J, Berger M, Saltzman J, Welcome M. Three-dimensional adaptive mesh refinement for hyperbolic conservation laws. SIAM J Sci Comput 1994;15(1):127–38.

[38] Bank RE, Sherman AH, Weiser A. Some refinement algorithms and data structures for regular local mesh refinement. In: Scientific Computing. Applications of Mathematics and Computing to the Physical Sciences, vol. 1. 1983. p. 3–17.

[39] Shepherd JF, Dewey MW, Woodbury AC, Benzley SE, Staten ML, Owen SJ. Adaptive mesh coarsening for quadrilateral and hexahedral meshes. Finite Elem Anal Des 2010;46(1):17–32.

[40] Prasad NS, Narayanan S. Finite element analysis of temperature distribution during arc welding using adaptive grid technique. Weld J (USA) 1996;75(4):123.

[41] Runnemalm H, Hyun S. Three-dimensional welding analysis using an adaptive mesh scheme. Comput Methods Appl Mech Eng 2000;189(2):515–23.

[42] Denlinger ER, Heigel JC, Michaleris P. Residual stress and distortion modeling of electron beam direct manufacturing Ti-6Al-4V. Proc Inst Mech Eng, B J Eng Manuf 2014;229:1803–13.

[43] Yu G, Masubuchi K, Maekawa T, Patrikalakis NM. A finite element model for metal forming by laser line heating. In: Proceedings of the 10th international conference on computer applications in shipbuilding. ICCAS, vol. 99. 1999. p. 409–18.

[44] Zhang L, Reutzel E, Michaleris P. Finite element modeling discretization requirements for the laser forming process. Int J Mech Sci 2004;46(4):623–37.

[45] Hughes TJ. The finite element method: linear static and dynamic finite element analysis. Mineola, NY: Dover Publications, Inc.; 2000.

[46] Lee NS, Bathe KJ. Error indicators and adaptive remeshing in large deformation finite element analysis. Finite Elem Anal Des 1994;16(2):99–139.

[47] Goldak J, Chakravarti A, Bibby M. A new finite element model for welding heat sources. Metall Trans B 1984;15(2):299–305.

11

Mitigation of Distortion in Large Additive Manufacturing Parts*

Erik R. Denlinger

Product Development Group, Autodesk Inc., State College, PA, United States

11.1 INTRODUCTION

Additive manufacturing (AM) processes allow for the construction of parts directly from a digital drawing file without the retooling cost associated with casting and forging. Wire or powder is melted onto a substrate by a laser or electron beam and allowed to cool and solidify to form a fully dense geometry built up on a layer by layer basis. The large thermal gradients caused by the deposition process often lead to significant workpiece distortion, especially in large builds, taking the part out of tolerance. In order to combat the issue of process-induced distortion, and in order to make AM processes useful in industry applications, techniques to mitigate distortion in large deposited parts must be developed.

In order to reduce distortion the appropriate distortion mode must be identified as first defined by Masubuchi [1]. The out-of-plane distortion modes include angular, buckling, and longitudinal bending distortion. Angular and buckling distortion are caused by similar mechanisms. Angular distortion is caused by transverse shrinkage in the deposition region, while buckling occurs when residual stress caused by longitudinal shrinkage exceeds the workpiece critical buckling strength. Both angular and buckling distortion are less common in AM than in welding as the substrates used are generally thicker than weld panels. Of the 3 possible out-of-plane modes, longitudinal bending is of primary concern in AM processes. The longitudinal bending distortion is caused by the contraction of the molten material after heating and deposition. The progression of substrate distortion during deposition is illustrated in Figure 11.1. When the molten material is applied the thermal expansion of the top of the substrate causes bending and plastic deformation, shown in Figure 11.1A. The final longitudinal bending of the substrate, displayed in Figure 11.1B, is caused by the cooling and contraction of the deposited material. When depositing large parts the problem of longitudinal bending

---

* This chapter is based upon the original work: Denlinger, Erik R., and Pan Michaleris. "Mitigation of distortion in large additive manufacturing parts." Proceedings of the Institution of Mechanical Engineers, Part B: Journal of Engineering Manufacture (2015): 983–993.

Thermo-Mechanical Modeling of Additive Manufacturing
DOI: 10.1016/B978-0-12-811820-7.00013-6

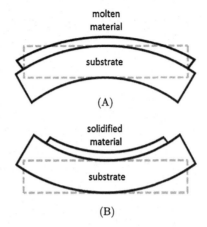

FIGURE 11.1 Progression of the substrate distortion throughout the deposition process. The undeformed and deformed substrates are illustrated by dashed and solid lines, respectively: (A) workpiece after the deposition of molten material and (B) workpiece after the cooling of deposited material.

can be exacerbated by the use of longer longitudinal deposition passes commonly used as a means to reduce processing time [2]. Several methods to reduce distortion incurred during welding and AM processes have been previously investigated.

Distortion mitigation techniques used in AM originate from research performed on a similar process, multi-pass welding. In welding research, finite element modeling (FEM) is commonly used to assess the effectiveness of distortion mitigation strategies while avoiding costly trial and error iterations. FEM development in welding dates back several decades and has focused on predicting both thermal and mechanical response of welded panels [3–8]. Weld research has shown reducing the heat input [9], balancing the residual stress to minimize the bending moment [9], and creating a temperature difference between parts to be welded (known as *transient differential heating*) [10] to be effective in reducing longitudinal bending distortion levels. AM differs from multi-pass welding in that AM involves the addition of large volumes of material, resulting in a larger number of deposition passes and longer processing times.

Researchers investigating distortion mitigation techniques in AM have shown that altering the laser scanning pattern can reduce distortion, with shorter deposition passes resulting in lower distortion levels [11–14]. However, shorter scanning patterns add processing time by requiring a greater number of deposition passes to add the same amount of material. This problem becomes more significant for large parts with a greater number of deposition passes. Residual stress may be reduced by preheating the substrate and holding it at a high bulk temperature [15–17] or by heating the deposition region immediately prior to deposition (*localized preheating*) [18]. Bulk substrate heating is only feasible for small workpieces as it is not practical to hold a large substrate at a high temperature for a long period of time, while localized preheating requires modifications to the laser or electron beam deposition system. Industry applications are frequently focused on large workpieces, but all of the distortion mitigation techniques thus far investigated have only been shown to be effective on models of small parts.

This work investigates 3 new distortion mitigation techniques, useful for reducing the longitudinal bending distortion mode in large AM parts. The first strategy involves applying heat to the workpiece substrate in an attempt to straighten it after deposition. The subsequent techniques involve depositing equal material on each side of a substrate in order to balance the bending moment about the neutral axis of the workpiece. The added deposition passes used to balance the bending moment are referred to as *balancing passes*, with the deposition passes needed to construct the actual part geometry referred to as *build passes*. The balancing passes deposit sacrificial material that will be machined away in a post-process. The second strategy examines the effect of depositing the build passes consecutively after the completion of all balancing passes is investigated. The third one investigates the possibility of depositing a balancing layer after each build layer is considered. The effectiveness of the techniques is investigated using a small FE model. The techniques found to be most successful on the small model are then applied to the manufacture of a large Ti-6Al-4V electron beam deposited part.

11.2 EVALUATION OF DISTORTION MITIGATION TECHNIQUES

The feasibility of applying heat to straighten a substrate or adding additional deposited material to balance the bending moment acting on the substrate is first investigated using differing deposition strategies on small models. Significant computation time can be saved by using small finite element models to predict the effectiveness of different distortion mitigation strategies on large parts when compared with simulating the actual large workpieces.

A brief overview of the model is first provided. Deposition cases are outlined which allow for the investigation of the aforementioned mitigation techniques. The deposition strategies that successfully achieve significant distortion mitigation on the small models will be applied to the manufacture of an actual large part.

11.2.1 Electron Beam Deposition Simulation

A 3D thermo-elasto-plastic analysis is performed to predict the effectiveness of the presented distortion mitigation techniques. The results of the thermal simulation are imported as a load file into the mechanical analysis, which does not affect the thermal analysis due to the fact that the two are weakly coupled. Electron beam deposition is modeled, as the process is commonly used to deposit large parts due to its ability to quickly deposit large amounts of bulk material. A detailed validation of the model is found in reference [19].

11.2.2 Numerical Model

A 203.2 mm long × 28.6 mm tall × 12.7 mm wide wall is constructed on the topside of the substrate and a 203.2 mm × 6.35 mm × 12.7 mm wall is built up on the backside of the substrate, as large and complex industry parts commonly require material to be placed on both sides of the substrate. Also, the part investigated in Section 4 possesses 9 layers and 2 layers on the top and bottom of the substrate, respectively. The depositions are made upon a 254 mm long × 101.6 mm wide × 12.7 mm thick substrate using unidirectional longitudinal

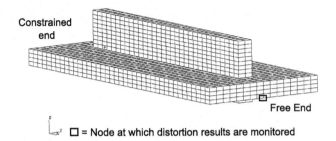

FIGURE 11.2   Mesh showing the node observed to monitor distortion.

TABLE 11.1   Distortion results for the small model

Case	Description
1	baseline
2	topside heating
3	sequential balancing layers
4	alternating balancing layers

passes. A cooling time of 1200 s occurs between the deposition of each layer. On large parts long cooling times are typical because each deposition layer can be time consuming, allowing the deposited material to cool significantly as the layer is completed. All meshes are generated using Patran 2012 by MSC. The thermal and mechanical analyses are performed using Netfabb Simulation by Autodesk Inc.

The addition of deposited material is simulated using the quiet element approach where all elements begin as part of the simulation, however elements belonging to the deposited material are given material properties such that they do not affect the analysis until contacted by the heat source. Surface radiation is applied to all free surfaces, including those on the evolving surface between active and quiet elements. The heat source has a power P of 8 kW and moves at a speed v_w of 12.7 mm/s. After the completion of the build, the parts cool to room temperature. The model is mechanically constrained as cantilevered allowing distortion to be monitored at a node on the free end of the substrate, as shown in Figure 11.2, to compare the different deposition strategies.

11.2.3 Deposition Strategies

Four cases are studied to determine the effectiveness of several mitigation strategies. Straightening the substrate by applying only heat and depositing additional material to balance the bending moment are compared to a baseline case. For actual builds, in the cases where additional material is deposited, the extra material is sacrificial and would be machined away post-process. A detailed description of each case is provided and the cases are summarized in Table 11.1.

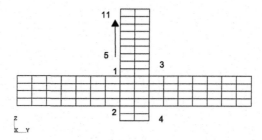

FIGURE 11.3 Baseline deposition pattern, Case 1.

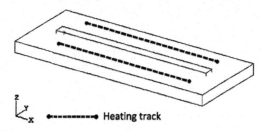

FIGURE 11.4 Topside heating with no sacrificial material added, Case 2.

Case 1: Baseline With Topside Deposition Only

Figure 11.3 shows the baseline case, Case 1. Nine layers and 2 layers are deposited on the topside and backside of the substrate respectively, as labeled in Figure 11.3. The mesh is comprised of 3264 elements and 4574 nodes. No attempted is made to reduce distortion accumulated during the deposition of material.

Case 2: Topside Heating After Each Topside Layer

Figure 11.4 shows Case 2 which applies topside heating in an attempt to lessen distortion of the part without adding additional material. The mesh is comprised of 3264 elements and 4574 nodes. After the deposition of each unbalanced deposition layer on the topside of the substrate 2 heating passes are performed each 203.2 mm long. The longitudinal bending distortion caused by the thermal expansion of the top of the substrate is intended to counter the distortion caused by the contraction of the molten material.

Case 3: Backside Deposition After Topside Layers Deposited

Case 3, whose deposition pattern is seen in Figure 11.5, uses consecutive balancing layers on the backside of the substrate to balance the build layers. The first 2 build layers on the topside of the substrate and the 2 build layers on the backside of the substrate are deposited in alternating fashion. Then the remaining 7 build layers on the topside of the substrate are deposited sequentially followed by 7 consecutive balancing layers on the backside of the substrate. The mesh has 3712 elements and 5267 nodes.

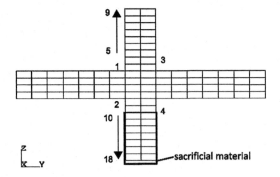

FIGURE 11.5   Sequential deposition pattern, Case 3.

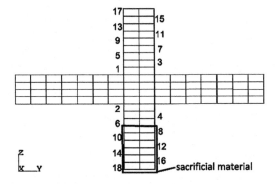

FIGURE 11.6   Alternating deposition pattern, Case 4.

Case 4: Backside Deposition After Each Topside Layer

Case 4, shown in Figure 11.6, examines the possibility of mitigating distortion by depositing a balancing layer on the backside of the substrate after each build layer on the topside of the substrate in order to continually balance the bending moment about the neutral axis of the workpiece throughout the build.

The end result is a workpiece with a 9 layer high wall on each side of the substrate. When the balancing layers cool after deposition, the bending moment caused by the contraction of the molten material of the balancing layers will equal the bending moment caused by the contraction of the molten material making up the build layers and thus may help to straighten the substrate. The mesh consists of 3712 elements and 5267 nodes.

11.2.4 Small Model Results

Figure 11.7 plots the results of Case 1, the baseline case, to illustrate how distortion is being quantified. In each case the maximum substrate distortion occurs at the free end of the cantilevered substrate. Table 11.2 summarizes the results of the simulations.

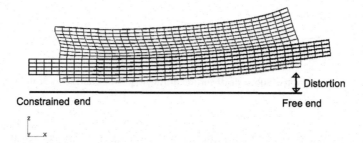

FIGURE 11.7   Final distortion for the baseline case, Case 1.

TABLE 11.2   Distortion results for the small models

Case	Description	Final Distortion (mm)	% Decrease
1	baseline	3.32	–
2	topside heating	2.04	38.9
3	sequential balancing layers	1.17	64.8
4	alternating balancing layers	0.06	98.2

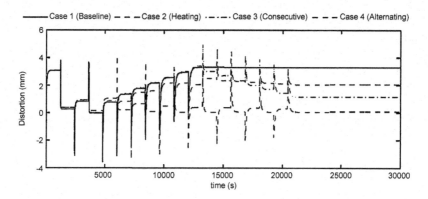

FIGURE 11.8   In-situ distortion for Cases 1–4.

For Case 1, the baseline where 9 and 2 layers are deposited on the topside and backside of the substrate respectively, it can be seen in Figure 11.8 that nearly no distortion has been accumulated after the first 4 deposition layers (around 4900 s) as the bending moment remains balanced about the neutral axis of the substrate. The remaining 7 deposition layers on the topside of the substrate cause a final distortion of 3.32 mm.

Applying topside heating to the model in Case 2 results in a final distortion of 2.04 mm, a 38.9% decrease compared with the baseline.

The in-situ distortion results from Case 3, applying sequential balancing layers, shows that distortion accumulates in the build until 12000 s when balancing layers begin to reduce it. The balancing layers are unable to mitigate all distortion and result in a final distortion of 1.17 mm which is a 64.8% reduction of distortion relative to the baseline case.

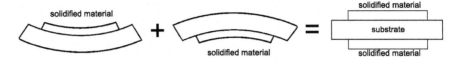

FIGURE 11.9 Evolution of substrate distortion during deposition and solidification leading to reduced distortion of the workpiece.

Case 4, using alternating balancing layers, can be seen to accumulate little distortion throughout the build. The balancing of the bending moment after each deposition layer results in a final distortion of 0.06 mm, representing a 98.2% reduction in distortion compared with Case 1, the baseline case.

The results from the small model suggest that balancing the bending moment about the substrate by adding material is the most capable method of those explored for reducing distortion in AM workpieces. The method yields superior results to applying only heating.

The distortion mitigation is achieved when the deposited balancing layers cool from their molten state and shrink to form a fully dense deposition, essentially canceling out the distortion caused by the build layers as illustrated in Figure 11.9.

The observed accumulation of distortion in the models indicates that the sequence in which the balancing layers are added is important. The balancing layers in Case 3, which are deposited sequentially, result in significant distortion mitigation, however, the balancing layers in Case 4, which are deposited in alternating fashion, eliminate nearly all distortion attributed to the deposition of the build layers. The alternating balancing layers in Case 4 eliminate significantly more distortion than their sequentially deposited counterparts in Case 3. This result suggests that depositing the build layers and balancing layers in an alternating manner can be used to eliminate virtually all longitudinal bending distortion.

11.3 MITIGATION TECHNIQUES APPLIED ON A LARGE PART

To further investigate the effectiveness of the distortion mitigation strategies presented, a large electron beam deposited build is constructed twice to study if the most successful strategies from the small models result in similar distortion reduction on a larger scale.

11.3.1 Experimental Procedure

An electron beam freeform fabrication system is used to deposit 9.5 mm diameter Ti-6Al-4V wire feedstock material, at a rate of 0.85 mm/s, in a vacuum chamber, to form the workpiece shown in Figure 7 in Chapter 11. The AM system used is the Sciaky VX-300, which welds in a vacuum in the range of 10^{-4} to 10^{-5} Torr. The work envelope is approximately $5.8 \times 1.2 \times 1.2 \, m^3$ in volume. The electron beam power is varied from 8 kW to 10 kW in order to control the melt pool size. The largest builds are deposited 80 layers high, with a total deposition layer count of 107. A Ti-6Al-4V plate 3810 mm long, 457 mm wide, and 25.4 mm thick is used as a substrate. Two parts, Part A and Part B, are deposited on each

FIGURE 11.10   Fixture used to constrain the substrate during deposition.

TABLE 11.3   Case descriptions for the large part

Case	Description
L1	baseline
L2	alternating balancing layers
L3	sequential balancing layers

substrate during the build. Two builds are performed, Build 1 and Build 2, allowing for a total of 4 parts to be manufactured. Post-process scan results, taken by Neomek Inc, allow for the workpiece distortion levels to be quantified.

Figure 11.10 shows the test fixture used. The substrate is placed on the fixture and held in place by 40 evenly spaced clamps. The clamps have a spring constant of 22.5 N/mm, allowing for in-situ distortion of the substrate. The fixture can rotate to allow for deposition on both sides of the substrate.

11.3.2 Deposition Cases

Due to the effectiveness of the distortion mitigation strategies applied on Cases 3 and 4 these same strategies are implemented on a large part. The new Cases, L2 and L3, will be deposited using scan patterns which deposit balancing layers alternatingly and sequentially, respectively and compared to a new baseline deposition Case L1. These cases are summarized in Table 11.3.

Case L1

Figure 11.11 shows Case L1 which is applied to Build 1 and excludes the use of balancing layers. The deposition strategy is the same as that applied in Case 1 and is applied to both parts deposited on the substrate of Build 1 making Part A and Part B identical. Case L1 is the baseline case used to determine the effectiveness of the distortion mitigation strategies.

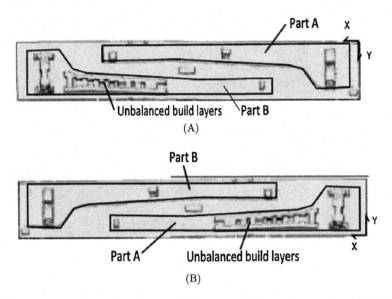

FIGURE 11.11 Schematic diagram of the large build illustrating both parts deposited on build 1 (case L1): (A) top view and (B) bottom view.

The layers on Case L1 which are responsible for the majority of the longitudinal bending distortion of the workpiece have been identified in previous finite element modeling work by the authors [2] and are labeled in Figure 11.11.

Case L2 and Case L3

Case L2 employs a strategy that is identical to Case 4, using alternating build layers and balancing layers. The strategy is applied to Part A of Build 2 as shown in Figure 11.12A. After the completion of each build layer the part is physically turned over to allow for the deposition of a balancing layer, and then returned to its initial position to resume the deposition of the next build layer.

The deposition strategy used for Case L3 is analogous to that of Case 3 and applies sequential build layers, followed by sequential balancing layers. Figure 11.12B shows the additional layers added to Part B of Build 2. After all build layers have been added to the part, it is turned over to allow for the addition of the balancing layers. This strategy requires fewer rotations of the part and thus reduces processing time.

After the completion of the deposition of the 2 builds, post-process scan results are taken by Neomek, Inc to allow for the workpiece distortion levels to be quantified.

11.4 RESULTS AND DISCUSSION

The distortion results are plotted in Figure 11.13. Case L1, the unbalanced deposition applied to Build 1, accumulates a maximum of 37.2 mm of distortion on Part A and 32.6 mm on

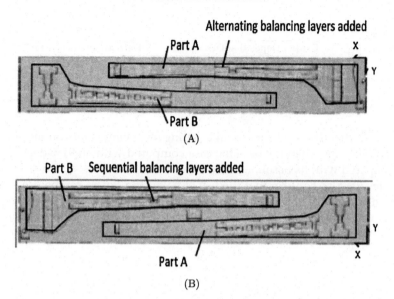

FIGURE 11.12 Schematic diagram of the large build illustrating both parts deposited on build 2 (cases L2 and L3): (A) top view of build 2 showing the added alternating deposition layers used to balance the part and (B) bottom view of build 2 showing the added sequential deposition layers used to balance the part.

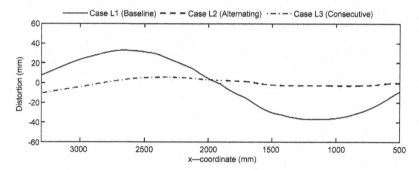

FIGURE 11.13 Comparison of distortion results before and after application of the distortion mitigation techniques on the targeted portion of the substrate.

Part B. Case L2 and Case L3 are implemented on Part A and Part B of Build 2, respectively. The Case L2 deposition pattern results in a maximum substrate distortion of 3.3 mm on Part A, representing an 91% decrease in distortion compared with the baseline case. Case L3 results in distortion levels as high as 10.2 mm on Part B. Case L3 has a maximum distortion 69% smaller than Case L1. Table 11.4 summarizes the results.

The application of the distortion mitigation strategies on the large part yields similar results as are found from the small models. The small model intended to represent a large workpiece, predicted a percent distortion mitigation of 65% when adding sequential balancing layers whereas the manufactured large part using sequential balancing layers saw a 69%

TABLE 11.4 Distortion results for the large part

Case	Deposition Description	% Decrease in Max Distortion
L1	unbalanced	-
L2	alternating	91
L3	sequential	69

reduction in distortion. The use of alternating balancing layers on the small model yielded a distortion 98% less than the baseline case compared with an 91% decrease seen on the large part compared with its baseline case.

The results also confirm that the use of balancing layers can be used to significantly mitigate distortion of large workpieces. Depositing the balancing layers sequentially after the deposition has been completed can be done to save processing time, however the greatest distortion mitigation is achieved when depositing a balancing layer after each build layer.

11.5 CONCLUSIONS

Several distortion mitigation strategies for AM parts have been presented to allow for a significant reduction in the longitudinal bending of a workpiece. The approaches involve using heating to straighten a substrate or depositing additional material to balance the bending moment about the neutral axis of the workpiece. The distortion mitigation strategies are well suited for large parts, as they do not require any modifications to the AM system and do not require the impractical heating of the entire substrate to limit thermal gradients.

The effectiveness of the strategies is demonstrated on a small FE model. This allows for the lengthy computation time associated with simulating large parts to be avoided. The small models show that adding additional material is a superior mitigation technique compared with applying only heating, thus the strategy is applied to the manufacture of large parts.

The distortion results of the large parts indicate that applying a balancing layer after each build layer yields favorable results when compared with depositing all balancing layers sequentially after the completion of the build layers and is capable of eliminating nearly all bending distortion. This consequence is in agreement with the prediction from the small models. The percent reduction in distortion on the large parts is found to be in close agreement with that calculated in the simulations. It is suggested that when designing a build plan an effort should be made to deposit roughly equal volume of material on each side of the substrate in order to minimize substrate distortion.

References

[1] Masubuchi K. Analysis of welded structures: residual stresses, distortion, and their consequences. New York: Pergamon Press; 1980.
[2] Denlinger E, Irwin J, Michaleris P. Thermo-mechanical modeling of additive manufacturing large parts. J Manuf Sci Eng 2014; Manuscript in review.
[3] Hibbitt HD, Marcal PV. A numerical, thermo-mechanical model for the welding and subsequent loading of a fabricated structure. Comput Struct 1973;3(5):1145–74.

[4] Argyris JH, Szimmat J, Willam KJ. Computational aspects of welding stress analysis. Comput Methods Appl Mech Eng 1982;33(1):635–65.

[5] Free J, Porter Goff R. Predicting residual stresses in multi-pass weldments with the finite element method. Comput Struct 1989;32(2):365–78.

[6] Tekriwal P, Mazumder J. Finite element analysis of three-dimensional transient heat transfer in GMA welding. Weld J 1988;67(5):150–6.

[7] Michaleris P, Tortorelli DA, Vidal CA. Analysis and optimization of weakly coupled thermoelastoplastic systems with applications to weldment design. Int J Numer Methods Eng 1995;38(8):1259–85.

[8] Lindgren LE, Runnemalm H, Näsström MO. Simulation of multipass welding of a thick plate. Int J Numer Methods Eng 1999;44(9):1301–16.

[9] Michaleris P. Minimization of welding distortion and buckling: modelling and implementation. WP, Woodhead Publishing; 2011.

[10] Deo M, Michaleris P. Mitigation of welding induced buckling distortion using transient thermal tensioning. Sci Technol Weld Join 2003;8(1):49–54.

[11] Nickel A, Barnett D, Prinz F. Thermal stresses and deposition patterns in layered manufacturing. Mater Sci Eng A 2001;317(1):59–64.

[12] Kruth J, Froyen L, Van Vaerenbergh J, Mercelis P, Rombouts M, Lauwers B. Selective laser melting of iron-based powder. J Mater Process Technol 2004;149(1):616–22.

[13] Dai K, Shaw L. Distortion minimization of laser-processed components through control of laser scanning patterns. Rapid Prototyping J 2002;8(5):270–6.

[14] Foroozmehr E, Kovacevic R. Effect of path planning on the laser powder deposition process: thermal and structural evaluation. Int J Adv Manuf Technol 2010;51(5–8):659–69.

[15] Vasinonta A, Beuth J, Griffith M. Process maps for controlling residual stress and melt pool size in laser-based SFF processes. In: Solid freeform fabrication proceedings. Proc. 2000 solid freeform fabrication symposium; 2000. p. 200–8.

[16] Jendrzejewski R, Śliwiński G, Krawczuk M, Ostachowicz W. Temperature and stress fields induced during laser cladding. Comput Struct 2004;82(7):653–8.

[17] Jendrzejewski R, Śliwiński G. Investigation of temperature and stress fields in laser cladded coatings. Appl Surf Sci 2007;254(4):921–5.

[18] Aggarangsi P, Beuth J. Localized preheating approaches for reducing residual stress in additive manufacturing. In: Proc SFF symp, Austin; 2006. p. 709–20.

[19] Denlinger E, Heigel J, Michaleris P. Residual stress and distortion modeling of electron beam direct manufacturing Ti-6Al-4V. Proc Inst Mech Eng, B J Eng Manuf 2014;229:1803–13.

THERMOMECHANICAL MODELING OF POWDER BED PROCESSES

Development and Numerical Verification of a Dynamic Adaptive Mesh Coarsening Strategy for Simulating Laser Power Bed Fusion Processes

Erik R. Denlinger

Product Development Group, Autodesk Inc., State College, PA, United States

12.1 INTRODUCTION

Laser Powder-Bed Fusion (LPBF) is an additive manufacturing process capable of producing net and near-net shape parts directly from a digital drawing file with dimensional tolerances of less than 0.1 mm [1]. Like other AM processes, such as Directed Energy Deposition (DED) and Electron Beam Additive Manufacturing (EBAM), the thermal history of the part will determine if the component develops defects, if the resulting microstructure is uniform and acceptable, and if the residual stress levels will take the part out of tolerance or cause failure. These issues can be addressed by applying finite element analysis (FEA), rather than through a costly experimental trial and error process. However, for LPBF, laser radii are typically in the range of 0.05 to 0.25 mm and scan speeds are in the range of 600 to 1300 mm/s [2–6]. These parameters impose strict spacial and temporal discretization requirements in the analyses, resulting in infeasible run times. In order to simulate the LPBF process, meshing strategies to reduce the number of degrees of freedom in the analyses are needed.

The basis of the FEA approach used for modeling AM processes extends back nearly 40 years into the welding literature as the two processes are similar. FEA has been extensively applied to predict thermal history of welding processes [7–16]. State of the art summaries on the development of weld modeling are available in references [17–19].

Thermo-Mechanical Modeling of Additive Manufacturing
DOI: 10.1016/B978-0-12-811820-7.00015-X

The research on weld process modeling has been extended to DED processes. Several models are available for predicting the thermal response of DED processes [20–35]. The primary difference in process modeling between the LPBF process and the aforementioned DED processes is that the LPBF process uses a laser with a relatively small spot size. The small heat source size imposes taxing spacial and temporal discretization requirements due to the fact that the number of elements required in the mesh and the acceptable size of the time steps directly depends on the spot size of the heat source [36].

Several researchers have developed thermal models for the LPBF process, focusing on small geometries. Roberts et al. validated a thermal model of the LPBF deposition of 0.15 mm^3 of titanium powder [6]. Li et al. developed and validated a temperature dependent microstructural model for single layer 2 mm^2 titanium builds [37]. The thermal model developed in reference [38] was used to simulated the deposition of a 1 mm^3 volume of nickel. The noted LPBF thermal models provide insight into the thermal cycles experienced during the manufacturing process however it would be infeasible to simulate larger deposition volumes as the problems would become too computationally expensive.

In order to reduce the computational expense of the simulations, past work has focused on reducing the magnitude of the temporal part of the problem. Wanxie et al. developed a scheme that allows for larger time increments to be taken for larger elements [39]. Irwin developed an elongated model for the heat source for AM processes which allows for a significant reduction in time steps needed to complete the analysis. These approaches increase the feasibility of simulating moving-source thermal problems but do not address the issue of large numbers of DOFs present in the analysis.

Two strategies that have been implemented to reduce the size of the spacial problem include the use of nonconforming meshes and adaptive meshes. Kolossov et al. presented a moving source thermal model for the Selective Laser Sintering (SLS) process and simulated the sintering 2 mm^3 of material [21]. A nonconforming meshing was utilized to allow for the mesh to be quickly coarsened in areas away from the deposition region. Adaptive meshing has been used for 2D weld simulations [40]. Runnemalm and Hyunn developed a 3D Hex element adaptive scheme for welding but memory demands of the basemesh were too large to access the accuracy of the approach. To improve the speed of computation for LPBF processing, Zeng et al. applied a dynamic mesh coarsening algorithm to refine the mesh around the heat source at each time step in the analysis [41] thus reducing the number of degrees of freedom (DOFs) in the analysis. No modeling strategy presented allows for the simulation of large scan areas. To further reduce run times for LPBF analyses, an adaptive algorithm that offers greater computational time savings is needed.

In this chapter, a mesh coarsening strategy is demonstrated which significantly reduces the spacial discretization problem for moving source AM simulations. The strategy investigated is dynamic adaptive meshing, which keeps a fine mesh around the heat source and automatically coarsens the mesh away from the moving source at each time step. The accuracy of the methods is verified by comparing the calculated thermal results with the results of a static conforming mesh. The required computational time for the analysis is compared to run times for the same geometry using a static conforming mesh and a static nonconforming mesh in order to demonstrate the potential time savings introduced by the use of dynamic mesh coarsening. Criteria for mesh coarsening are investigated.

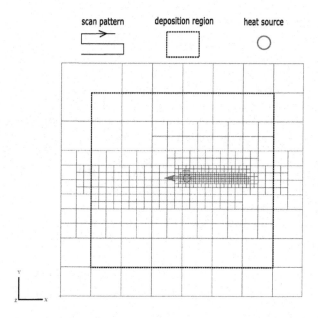

scan pattern   deposition region   heat source

FIGURE 12.1   Analysis using dynamic adaptive meshing.

12.2 MODELING APPROACH

A transient thermal analysis is performed followed by a quasi-static elasto-plastic mechanical analysis. The analyses are weakly coupled because the strain energy is small compared with the thermal energy introduced by the heat source [42]. The modeling was completed using Netfabb Simulation, following the methodology detailed in Chapter 2.

12.3 MESHING STRATEGIES

The results and run times achieved using the dynamic adaptive meshing strategy are evaluated against two more common mesh types, static nonconforming meshing and static conforming meshing. A description of each is provided.

12.3.1 Dynamic Adaptive Mesh Implementation

A dynamic mesh coarsening strategy is developed to significantly decrease the size of the spacial discretization requirement. Figure 12.1 illustrates the concept of dynamic adaptive meshing. The necessary computational time is reduced by keeping a fine mesh around the heat source and coarsening regions of the mesh that do not require fine elements at each time step.

12.3.1.1 Dependent Nodes

As the elements in the thermal analysis using dynamic adaptivity are coarsened the mesh may become nonconforming, i.e., the node of an element may not be shared by an adjacent element. The node will instead fall on the edge or face of the adjacent element and become a dependent (condensed) node. The dependent nodes are accounted for in the analysis by applying condensation and recovery [43]. Any associated degrees of freedom (DOFs) $\mathbf{U}$ for the dependent node must be constrained to free (conforming) nodes. At each time-step I the incremental solution $\delta\mathbf{U}$ is calculated as follows:

$$\delta\mathbf{U} = -\left[\frac{d\mathbf{R}}{d\mathbf{U}}(\mathbf{U}^I)\right]^{-1}\mathbf{R}(\mathbf{U}^I) \tag{12.1}$$

where $\mathbf{R}$ is the residual.

The independent (retained) DOFs $\mathbf{U_r}$ and dependent (condensed) DOFs $\mathbf{U_c}$ can be grouped allowing for (12.1) to be written as:

$$\begin{Bmatrix} \delta\mathbf{U_r} \\ \delta\mathbf{U_c} \end{Bmatrix} = \begin{bmatrix} \frac{d\mathbf{R}}{d\mathbf{U_r}}(\mathbf{U_r}^I) & \frac{d\mathbf{R}}{d\mathbf{U_r}}(\mathbf{U_c}^I) \\ \frac{d\mathbf{R}}{d\mathbf{U_c}}(\mathbf{U_r}^I) & \frac{d\mathbf{R}}{d\mathbf{U_c}}(\mathbf{U_c}^I) \end{bmatrix}^{-1} \begin{Bmatrix} \mathbf{R}(\mathbf{U_r^I}) \\ \mathbf{R}(\mathbf{U_c^I}) \end{Bmatrix} \tag{12.2}$$

The constraint equations have the following form:

$$U_c = C_r \sum_{k=1}^{n} U_k \tag{12.3}$$

where C_r is a constraint coefficient and n is the number of nodes that U_c is dependent upon. If an element's node lies on an adjacent element's edge, C_r and n equal $\frac{1}{2}$ and 2, respectively. If an element's node lies on an adjacent element's face, C_r and n equal $\frac{1}{4}$ and 4, respectively. By enforcing the constraints the system can be solved by including only the retained degrees of freedom:

$$\{\delta\mathbf{U_r}\} = \left[\frac{d\mathbf{R}}{d\mathbf{U_r}}(\mathbf{U_r}^I) - \frac{d\mathbf{R}}{d\mathbf{U_c}}(\mathbf{U_r}^I)\mathbf{C_r} + \mathbf{C_r^T}\frac{d\mathbf{R}}{d\mathbf{U_c}}(\mathbf{U_c}^I)\mathbf{C_r}\right]^{-1}\{\mathbf{U_r} - \mathbf{C_r^T}\mathbf{U_c}\} \tag{12.4}$$

The solution of the condensed nodes is then determined by enforcing the constraint equations.

12.3.1.2 Mesh Coarsening Criterion

When applying dynamic adaptivity, an element coarsening criterion must be established in order to know when it is permissible to coarsen a region of the mesh. If the mesh is coarsened too aggressively undesirable amounts of artificial energy $E_{artificial}$ can be added into the system. Figure 12.2 is a 1D example that illustrates this potential issue.

The original fine mesh contains 2 elements (Element 1 and Element 2) and 3 nodes. The elements are then combined to represent a coarser mesh containing 1 element (Element 3) and 2 nodes. The coarsening introduces artificial energy E_a which can be quantified as:

$$E_{artificial} = E_{coarse} - E_{fine} \tag{12.5}$$

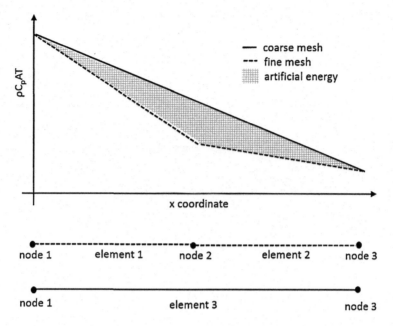

FIGURE 12.2 1D mesh coarsening example.

where E_{fine} and E_{coarse} is the energy stored in the original fine mesh and the coarsened mesh, respectively. The energy E stored in an element number j with an area A is:

$$E^j = \rho C_p A \int_x T^j(x)dx \tag{12.6}$$

From this example it can be seen that if the temperature difference between the nodes in each element equals zero then the artificial energy $E_{artificial}$ added into the system will also equal zero. For elements containing more than two nodes the coarsening criterion for each element in the mesh is as follows:

$$T_{max} - T_{min} <= \Delta T_c \tag{12.7}$$

where T_{max} and T_{min} are the maximum and minimum nodal temperatures of the element at particular time step, respectively. ΔT_c is the maximum allowable temperature difference between any two nodes in the element in order to allow coarsening. Setting $\Delta T_c = 0$ should results in no energy error when compared with a static mesh. When an element satisfies this criterion it is eligible to be combined with neighboring elements and may advance 1 level of adaptivity per time step. Levels of adaptivity are illustrated in Figure 12.3.

A significant limitation of moving source dynamic mesh coarsening is its inability to adequately model the mechanical response of an AM build. In thermal analyses the temperature dissipates throughout the part, thermal gradients across the linear Hex8 elements will reach a sufficiently small level as to allow element coarsening without incurring large errors in the

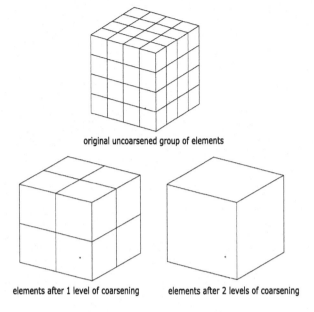

original uncoarsened group of elements

elements after 1 level of coarsening   elements after 2 levels of coarsening

FIGURE 12.3   Illustration of levels of coarsening.

predicted temperature. Unfortunately, the same concept cannot be applied for mechanical analyses. This is due to the fact that the plastic strain field is inherently permanent and will not dissipate. Coarsening the mesh away from the moving heat source will result in a loss of resolution in the plastic strain field. Thus, in this work, only thermal analyses are investigated.

12.3.2 Static Nonconforming Mesh Analysis

The first type of meshing strategy that the dynamic adaptive scheme is compared to is static nonconforming meshing. Figure 12.4 shows an example of a static nonconforming mesh. The mesh stays constant throughout the entire analysis but can be quickly coarsened away from the deposition region by enforcing the constraint relations presented in the previous chapter. A static nonconforming mesh will contain more nodes and elements than a mesh used for a dynamic adaptive analysis but potentially far fewer nodes and elements than required for a static conforming mesh.

12.3.3 Static Conforming Mesh Analysis

Figure 12.5 displays a static conforming mesh. Every node in the mesh must share a node with an adjacent element. No constraint equations are necessary. The ability to coarsen the mesh away from the deposition region is limited.

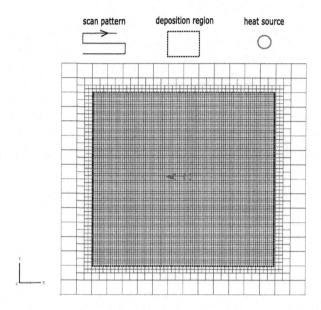

FIGURE 12.4 Analysis using a static nonconforming mesh.

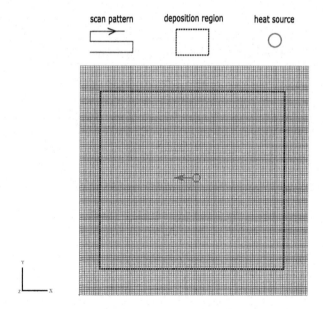

FIGURE 12.5 Analysis using a static conforming mesh.

TABLE 12.1   As-used constant material properties and processing conditions

Parameter	Value
Ambient temperature T_∞ [°C]	25
Convection coefficient h [W/m^2/°C]	10
Density ρ [kg/m^3]	4430
Emissivity ε	0.25
Hatch spacing [mm]	0.10
Laser absorptivity η	0.35
Laser diameter mm	0.10
Laser power [W] P	140
Laser scan speed [mm/s] v	1250
Poisson's ratio v	0.34

12.4 VERIFICATION

12.4.1 Numerical Implementation

In order to verify the accuracy and efficiency of the presented meshing strategies, simulations are performed using the same geometry and processing conditions for dynamic adaptive meshing, static nonconforming meshing, and static conforming meshing. A 3 × 3 mm^2 deposition region is heated on a 4 × 4 × 4 mm^3 substrate. The scan direction alternates between the positive and negative x direction. The analyses are performed in Netfabb Simulation. Each node has 1 DOF in the thermal analysis (temperature). The finest elements in each mesh are 0.03125 × 0.03125 mm^2 in the x–y plane and 0.05 mm in the z direction to simulate the melting of a 0.05 mm thick powder layer. The spacial discretization was chosen based on the known requirements needed in the Netfabb Simulation solver from the literature. In addition a 3-step spacial mesh convergence study was performed to insure that the mesh is sufficiently fine. The results converge between 1–2 elements per laser radius. The analyses are done in a series of time steps with the current time step taking the solution at the previous time step as an initial condition. The temporal discretization requirement for laser based AM process simulations in Netfabb Simulation is known from previous work [20,28,30]. The temperature dependent thermal and mechanical properties for Ti-6Al-4V were taken from [27]. Table 12.1 lists the as-used constant material properties and processing conditions for the analyses.

The laser heat source Q is modeled using the double ellipsoid model described in Chapter 2.

12.4.2 Assessment of Accuracy

The thermal results from the simulations using a static conforming, static nonconforming, and moving source adaptive mesh are compared along lines in the middle of the substrate. Figure 12.6 shows the lines along which the results are compared in the x–y plane.

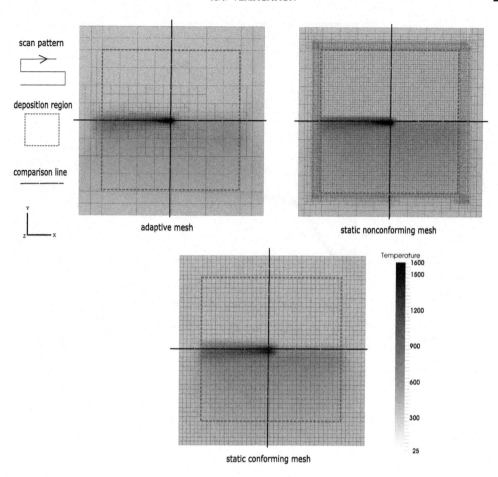

FIGURE 12.6 Comparison points used for verification of results.

Temperature versus x and y location for each mesh type is plotted in Figure 12.7A and Figure 12.7B respectively. It can be seen that both the nonconforming static mesh and the dynamic adaptive mesh predict temperatures that are in close agreement with the benchmark static conforming mesh. In order to quantify the errors $\delta_\%$ in the x direction $\delta_{x\%}$ and the y direction $\delta_{y\%}$ associated with each meshing strategy the errors are averaged over the temperature profile:

$$\delta_\% = \frac{100}{n} \times \sum_{i=1}^{n} \left| \frac{(T)_i - (\hat{T})_i}{(T)_i - T_\infty} \right| \tag{12.8}$$

where T and $\hat{T}$ are the temperatures calculated from the static conforming mesh and one of the other meshing strategies, respectively. The value of n is equal to the number of nodes along the comparison line.

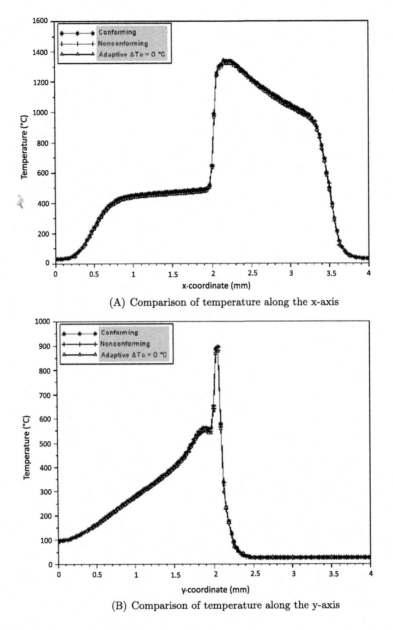

(A) Comparison of temperature along the x-axis

(B) Comparison of temperature along the y-axis

FIGURE 12.7 Verification results of the thermal analyses.

The calculated errors confirm the observations from Figure 12.7. The errors averaged over the deposition region for the static nonconforming mesh compared with the benchmark mesh in the x and y directions are equal to 2.99% and 3.04%, respectively. The errors associated with

TABLE 12.2 Comparison of accuracy and run times for each meshing strategy on a 3 mm × 3 mm deposition area

Mesh description	ΔT_c (°C)	$\delta_x \%$	$\delta_y \%$	4 core run time (s)	1 core run time (s)	% of static conforming run time	% of nonconforming run time
Static conforming	–	–	–	87548	350134	–	–
Static nonconforming	–	2.99	3.04	1962	7846	2.2	–
Moving adaptive	0	3.95	3.96	1256	5019	1.4	64
Moving adaptive	50	4.21	6.74	202	809	0.23	10
Moving adaptive	100	5.72	6.89	144	577	0.16	7.4
Moving adaptive	200	9.61	12.9	102	409	0.12	5.2

the dynamic meshing algorithm compared with static mesh are 3.95% and 3.96% along the x and y comparison lines, respectively. This indicates that the errors in the dynamic meshing simulation are mostly attributable to the use of the constraint equations.

12.5 RESULTS AND DISCUSSION

12.5.1 Effect of Coarsening Temperature on Dynamic Meshing Accuracy

To reduce the computation time required for the moving source adaptivity, the acceptable coarsening temperature difference ΔT_c may be increased. While in the previous section a value of 0 was chosen for ΔT_c to minimize errors, in many applications it may be acceptable to introduce small errors in order to save computation time. In order to explore the effect of increasing ΔT_c on the computation time and the accuracy of the simulation, additional simulations are performed varying ΔT_c from 0 to 200 °C in increments of 50 °C.

Figure 12.8 shows the temperature predictions resulting from the additional simulations. The comparisons are made along the same lines that are illustrated in Figure 12.6. As ΔT_c is increased, it can be seen that the artificial energy added into the system, that was illustrated in Figure 12.2, begins to manifest itself as an increased temperature prediction. As ΔT_c approaches 200 °C the increase in the temperature prediction becomes significant.

Table 12.2 summarizes the run times and errors associated with each analysis that was run with the 3 × 3 mm^2 deposition area. As in the previous section, errors were calculated by taking the static conforming mesh as a baseline. The use of a static nonconforming mesh, which introduces negligible error when compared with the static conforming mesh, decreases the run time by over 44 times. Dynamic adaptive meshing with no tolerance for temperature gradients across an element, with ΔT_c of 0, reduces the run time by nearly 70 times. The errors remain under 4%. However this is a strict and impractical use of moving adaptivity.

Raising ΔT_c to 50 °C significantly reduces the run time by twelvefold, with comparison to the noncomforming mesh, and 432 times faster than the nonadaptive mesh. The errors remain under 7%. As ΔT_c continues to increase to 100 and 200 °C, the errors become unacceptable.

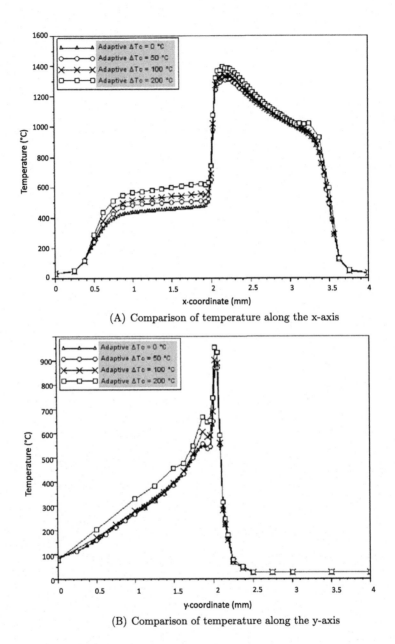

(A) Comparison of temperature along the x-axis

(B) Comparison of temperature along the y-axis

FIGURE 12.8   Temperature results for varying ΔT_c.

12.6 CONCLUSIONS

A dynamic adaptive meshing strategy has been developed for the rapid thermal modeling of LPBF processes. This method keeps a fine mesh in the melt pool region which has the steepest thermal gradients, while coarsening the rest of the mesh, updating the mesh each time step as the heat source moves. A mesh coarsening criterion is proposed, ΔT_C. This user specified value controls the degree to which the dynamically adaptive mesh is coarsened and describes the amount of allowable error using the constantly adaptive meshing strategy. To show the computational advantage of using dynamic meshing, simulations were completed using a perfectly static mesh, a static adaptive mesh that uses nonconforming elements to reduce the size of the element matrices, and 4 moving adaptive meshes using ΔT_C of 0, 50, 100, and 200 °C. Comparisons were made between run times and accuracy, using the static conforming mesh results as a comparative baseline. The static nonconforming mesh finished in 2.2% of the time the completely static mesh required, while introducing minimal error. It was shown that using ΔT_C of 100 or 200 °C introduced too much error to be reliable. Using ΔT_C of 0 reduced run time over the noncomforming mesh by almost 40%, with nearly equivalent accuracy. Of the 4 moving adaptive simulations attempted, the best compromise between speed and accuracy implemented a ΔT_C of 50. This case was 12 times faster than the nonconforming mesh and 432 times faster than the completely static mesh, while keeping the average error to about 5%. This study shows the utility and accuracy of using dynamically adaptive meshing for the thermal simulations of LPBF processes which may be used to create quick predictions of material properties or lack of fusion.

References

[1] Khaing M, Fuh J, Lu L. Direct metal laser sintering for rapid tooling: processing and characterisation of EOS parts. J Mater Process Technol 2001;113(1):269–72.

[2] Bugeda G, Cervera M, Lombera G. Numerical prediction of temperature and density distributions in selective laser sintering processes. Rapid Prototyping J 1999;5(1):21–6.

[3] Contuzzi N, Campanelli S, Ludovico A. 3D finite element analysis in the selective laser melting process. Int J Simul Model (IJSIMM) 2011;10(3).

[4] Morgan R, Sutcliffe C, O'neill W. Density analysis of direct metal laser re-melted 316L stainless steel cubic primitives. J Mater Sci 2004;39(4):1195–205.

[5] O'neill W, Sutcliffe C, Morgan R, Landsborough A, Hon K. Investigation on multi-layer direct metal laser sintering of 316L stainless steel powder beds. CIRP Ann-Manuf Technol 1999;48(1):151–4.

[6] Roberts I, Wang C, Esterlein R, Stanford M, Mynors D. A three-dimensional finite element analysis of the temperature field during laser melting of metal powders in additive layer manufacturing. Int J Mach Tools Manuf 2009;49(12):916–23.

[7] Hibbitt HD, Marcal PV. A numerical, thermo-mechanical model for the welding and subsequent loading of a fabricated structure. Comput Struct 1973;3(5):1145–74.

[8] Friedman E. Thermomechanical analysis of the welding process using the finite element method. J Press Vessel Technol 1975;97:206.

[9] Andersson B. Thermal stresses in a submerged-arc welded joint considering phase transformations. J Eng Mater Technol 1978;100 (United States).

[10] Argyris JH, Szimmat J, Willam KJ. Computational aspects of welding stress analysis. Comput Methods Appl Mech Eng 1982;33(1):635–65.

[11] Papazoglou V, Masubuchi K. Numerical analysis of thermal stresses during welding including phase transformation effects. J Press Vessel Technol 1982;104(3) (United States).

[12] Free JA, Porter Goff RF. Predicting residual stresses in multi-pass weldments with the finite element method. Comput Struct 1989;32(2):365–78.

[13] Tekriwal P, Mazumder J. Finite element analysis of three-dimensional transient heat transfer in GMA welding. Weld J 1988;67(5):150–6.

[14] Michaleris P, Tortorelli DA, Vidal CA. Analysis and optimization of weakly coupled thermoelastoplastic systems with applications to weldment design. Int J Numer Methods Eng 1995;38(8):1259–85.

[15] Lindgren LE, Runnemalm H, Näsström MO. Simulation of multipass welding of a thick plate. Int J Numer Methods Eng 1999;44(9):1301–16.

[16] Asadi M, Goldak JA. An integrated computational welding mechanics with direct-search optimization for mitigation of distortion in an aluminum bar using side heating. J Manuf Sci Eng 2014;136(1):011007.

[17] Lindgren LE. Finite element modeling and simulation of welding. Part 1: increased complexity. J Therm Stresses 2001;24(2):141–92.

[18] Lindgren LE. Finite element modeling and simulation of welding. Part 2: improved material modeling. J Therm Stresses 2001;24(3):195–231.

[19] Lindgren LE. Finite element modeling and simulation of welding. Part 3: efficiency and integration. J Therm Stresses 2001;24(4):305–34.

[20] Gouge MF, Heigel JC, Michaleris P, Palmer TA. Modeling forced convection in the thermal simulation of laser cladding processes. Int J Adv Manuf Technol 2015:1–14.

[21] Kolossov S, Boillat E, Glardon R, Fischer P, Locher M. 3D FE simulation for temperature evolution in the selective laser sintering process. Int J Mach Tools Manuf 2004;44(2):117–23.

[22] Peyre P, Aubry P, Fabbro R, Neveu R, Longuet A. Analytical and numerical modelling of the direct metal deposition laser process. J Phys D, Appl Phys 2008;41(2):025403.

[23] Qian L, Mei J, Liang J, Wu X. Influence of position and laser power on thermal history and microstructure of direct laser fabricated Ti-6Al-4V samples. Mater Sci Technol 2005;21(5):597–605.

[24] Shen N, Chou K. Thermal modeling of electron beam additive manufacturing process–powder sintering effects. In: Proceedings of the ASME 2012 international manufacturing science and engineering conference MSEC2012-7253; 2012. p. 1–9.

[25] Jamshidinia M, Kong F, Kovacevic R. Numerical modeling of heat distribution in the electron beam melting® of Ti-6Al-4V. J Manuf Sci Eng 2013;135(6):061010.

[26] Sammons PM, Bristow DA, Landers RG. Height dependent laser metal deposition process modeling. J Manuf Sci Eng 2013;135(5):054501.

[27] Denlinger ER, Irwin J, Michaleris P. Thermomechanical modeling of additive manufacturing large parts. J Manuf Sci Eng 2014;136(6):061007.

[28] Denlinger ER, Heigel JC, Michaleris P. Residual stress and distortion modeling of electron beam direct manufacturing Ti-6Al-4V. Proc Inst Mech Eng, B J Eng Manuf 2014;229:1803–13.

[29] Heigel J, Michaleris P, Palmer T. In situ monitoring and characterization of distortionduring laser cladding of Inconel® 625. J Mater Process Technol 2015.

[30] Heigel J, Michaleris P, Reutzel E. Thermo-mechanical model development and validation of directed energy deposition additive manufacturing of Ti-6Al-4V. Addit Manuf 2014.

[31] Anca A, Fachinotti VD, Escobar-Palafox G, Cardona A. Computational modelling of shaped metal deposition. Int J Numer Methods Eng 2011;85(1):84–106.

[32] Chiumenti M, Cervera M, Salmi A, Agelet de Saracibar C, Dialami N, Matsui K. Finite element modeling of multi-pass welding and shaped metal deposition processes. Comput Methods Appl Mech Eng 2010;199(37):2343–59.

[33] Lundbäck A, Lindgren LE. Modelling of metal deposition. Finite Elem Anal Des 2011;47(10):1169–77.

[34] Marimuthu S, Clark D, Allen J, Kamara A, Mativenga P, Li L, et al. Finite element modelling of substrate thermal distortion in direct laser additive manufacture of an aero-engine component. J Mech Eng Sci 2012.

[35] Mughal M, Fawad H, Mufti R. Three-dimensional finite-element modelling of deformation in weld-based rapid prototyping. J Mech Eng Sci 2006;220(6):875–85.

[36] Goldak J, Chakravarti A, Bibby M. A new finite element model for welding heat sources. Metall Trans B 1984;15(2):299–305.

[37] Li Y, Gu D. Thermal behavior during selective laser melting of commercially pure titanium powder: numerical simulation and experimental study. Addit Manuf 2014;1:99–109.

[38] Zhang D, Cai Q, Liu J, Zhang L, Li R. Select laser melting of W–Ni–Fe powders: simulation and experimental study. Int J Adv Manuf Technol 2010;51(5):649–58.

[39] Wanxie Z, Zhuang X, Zhu J. A self-adaptive time integration algorithm for solving partial differential equations. Appl Math Comput 1998;89(1):295–312.

[40] Prasad NS, Narayanan S. Finite element analysis of temperature distribution during arc welding using adaptive grid technique. Weld J (USA) 1996;75(4):123.

[41] Zeng K, Pal D, Gong H, Patil N, Stucker B. Comparison of 3DSIM thermal modelling of selective laser melting using new dynamic meshing method to ANSYS. Mater Sci Technol 2015;31(8):945–56.

[42] Zhang L, Reutzel E, Michaleris P. Finite element modeling discretization requirements for the laser forming process. Int J Mech Sci 2004;46(4):623–37.

[43] Cook RD, et al. Concepts and applications of finite element analysis. John Wiley & Sons; 2007.

13

Thermomechanical Model Development and In Situ Experimental Validation of the Laser Powder-Bed Fusion Process*

Erik R. Denlinger

Product Development Group, Autodesk Inc., State College, PA, United States

13.1 INTRODUCTION

The Laser Powder-bed Fusion (LPBF) additive manufacturing (AM) process allows for parts to be built on a layer by layer basis. Unlike other common AM processes, such as Laser Directed Energy Deposition (LDED) and Electron Beam Direct Manufacture (EBDM), the deposited powder is pre-placed rather than injected into the melt pool. The pre-placed metallic powder is traversed by the heat source causing the powder to melt and then solidify to form a fully dense geometry. After the material solidifies, the build platform is lowered and a recoater spreads a new layer of powder. The process is then repeated. The contraction of the molten material causes a buildup of residual stress which can lead to part failure by delamination from the buildplate, cracking, or recoater interference. The most common approach to addressing this issue is an expensive experimental trial and error process where parts are repeatedly manufactured until they are successfully built. The use of trial and error could be reduced or eliminated if there existed an experimentally validated thermomechanical finite element model (FEM) of the LPBF process to predict residual stress and distortion pre-process and allow for modifications to the build plan.

The basis of the FEM approach used for modeling AM processes extends back nearly 40 years into the welding literature as the two processes share many similarities. Researchers have been successful in using FEM to predict thermal history, distortion, and residual stress

---

* Should note that some content from this chapter was recently published: Denlinger, Erik R., et al. "Thermomechanical model development and in situ experimental validation of the Laser Powder-Bed Fusion Process." Additive Manufacturing (2017).

for welding processes [1–10]. State of the art summaries on the development of weld modeling are available in references [11–13].

The research on weld process modeling has been extended to Directed Energy Deposition (DED) processes. Simulations of AM processes are more computationally expensive than weld simulations due to the increased processing time and the need for elements to be continually added into the simulation to model the addition of deposited material. Several models are available for predicting the thermal response of DED processes [14–20]. Other models input the result of the thermal simulation into a mechanical analysis to predict distortion and residual stress [21–29].

Mechanical models of AM processes are typically validated using post-process measurements which provide no insight into the important in-process physics. To address this problem several researchers have validated models by using in-situ measurements. Lundbäck and Lindgren used an optical measurement system to experimentally validate the mechanical response of a Gas Tungsten Arc process [27]. Grum et al. [30] recorded in-situ strain using resistance measuring rosettes during laser cladding. Ocelik et al. used digital image correlation to measure distortion of single and multi-bead depositions [31]. Several works utilized a Laser Displacement Sensor (LDS) to record in-situ distortion for EBDM and LDED processes and used the measurements for thermomechanical model validation [24,23,22]. The use of in-situ measurement techniques provides more thorough model validation than can be achieved using only post-process measurements.

The primary difference in process modeling between the LPBF process and the aforementioned DED processes is that the LPBF process uses a laser with a relatively small spot size (as small as 70 μm). The small heat source size imposes taxing spacial and temporal discretization requirements due to the fact that the number of elements required in the mesh and the acceptable size of the time steps directly depends on the spot size of the heat source. Researchers have attempted to circumvent this issue by using analytical solutions [32] or by depositing entire deposition layers at once [33]. These approximations do not fully account for the plasticity introduced by a moving heat source.

Other researchers have focused solely on thermal modeling of small volumes of deposited material and have included a moving heat source in the simulations. Roberts et al. validated a thermal model of the LPBF deposition of 0.15 mm^3 of titanium powder [34]. Li et al. developed and validated a temperature dependent microstructural model for single layer 2 mm^2 titanium builds [35]. Kolossov et al. presented a moving source thermal model for the Selective Laser Sintering (SLS) process and simulated the sintering 2 mm^3 of material [15]. The thermal model developed in reference [36] was used to simulated the deposition of a 1 mm^3 volume of Nickel. The noted LPBF thermal models provide insight into the thermal cycles experienced during the manufacturing process however it would be infeasible to simulate larger deposition volumes as the problems would become computationally too expensive. To improve the speed of computation Zeng et al. applied a dynamic mesh coarsening algorithm to refine the mesh around the heat source at each time step in the analysis [37] thus reducing the number of degrees of freedom (DOFs) in the analysis. The approach reduced the computation time by a factor of 5 but was not extended to any large geometries or to solving mechanical problems. A mesh coarsening algorithm suitable for LPBF mechanical analyses is still needed.

A few studies have focused on modeling the mechanical response of the LPBF process. Matsumoto modeled the residual stress for single layer depositions [38]. Song et al. compared residual stress predictions to post-process X-ray diffraction strain measurements [39] for thin wall builds. Dai and Gu simulated 1.8 mm^3 of Copper deposition and validated their results with post-process density measurements [40]. No mechanical model available in the literature has been validated using in-situ distortion measurements. This is necessary to insure that the in-process physics of the LPBF are understood and adequately captured in the models.

The objective of this work is to develop a finite element model capable of accurately predicting the in-situ mechanical response of LPBF workpieces. Such a tool is needed to provide insight into the accumulation of residual stress, the effect of layer-to-layer interaction, and the suitability of the constitutive model for multi-layer builds. An improved understanding of the physics of the LPBF processes is a necessary step in order to eventually move toward full part-scale part modeling. Here, a thermomechanical simulation is run using a hybrid quiet/inactive element activation strategy [41]. As more layers are added into the thermal and mechanical analyses a mesh coarsening algorithm is utilized to merge the earlier deposition layers to further reduce the number of DOFs [21]. The effectiveness of the approach is demonstrated by simulating the mechanical response of a 38 layer 91 mm^3 volume build consisting of over 3,400 laser scan passes. In addition, the results are experimentally validated using in-situ distortion measurements from a deposition of the same geometry.

13.2 MODELING APPROACH

A detailed review of the finite element modeling of residual stress and distortion for welding and AM processes is given in reference [42]. In the present work Lagrangian reference frames are utilized. A transient thermal analysis is performed, followed by an elastoplastic mechanical analysis. The total Lagrangian formulation is chosen in this work due the fact that the substrate deposited upon is thin (under 1 mm) and may deform significantly. The process is weakly coupled, meaning that the mechanical results will have negligible effect on the thermal results. This is due to the fact that the energy from the heat source is very large when compared with the strain energy stored in the workpiece [43].

13.3 EXPERIMENTAL VALIDATION

The experimental in-situ distortion and temperature results presented in Ref. [44] are used in this work to validate the thermal and mechanical models. A brief explanation of the experimental setup is provided here and a detailed description can be found in Ref. [44].

13.3.1 Processing Parameters

The Inconel® 718 depositions were performed using the EOSINT M280 LPBF machine. The AM system uses a 4LR-400-SM-EOS laser. The part geometry and dimensions are shown in Figure 13.1. The hatch pattern rotates by a constant amount when the build progresses to a new layer. The processing parameters are listed in Table 13.1.

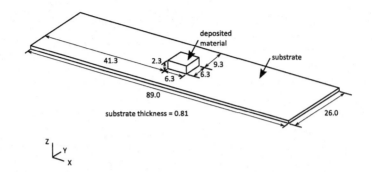

FIGURE 13.1 Part geometry (units: mm).

TABLE 13.1 Processing parameters

Parameter	Value
Average layer thickness [μm]	60
Hatch rotation angle [°]	67
Hatch spacing [μm]	110
Laser power P [W]	280
Laser scan speed v [mm/s]	960
Powder diameter D_p [μm]	30
Laser wavelength [nm]	1060–1100

13.3.2 Distortion and Temperature Measurements

The experimental setup is shown in Figure 13.2. The setup allows for in-situ distortion measurements to be taken using a differential variable reluctance transducer (DVRT) with an accuracy of ±1.5 μm. When the laser is on the top of the substrate, it expands upon heating, causing its upward bow. Then, upon cooling, the molten material contracts and causes a downward bow. These cycles are captured by the DVRT. In addition to the in-situ distortion measurements, in-situ distortion measurements were taken using K-Type thermocouples with an accuracy of ±2.2 °C. Figure 13.3 shows the measurement locations.

13.4 NUMERICAL IMPLEMENTATION

13.4.1 Solution Method

The thermal and elastoplastic analyses are performed in Netfabb Simulation by Autodesk Inc. The analyses are done in a series of time steps with the current time step taking the solution at the previous time step as an initial condition. The temporal discretization requirement for laser or electron beam based AM process simulations in Netfabb Simulation is known from previous work [24,14,22]. The modeling methodology used in this work follows that described in Chapter 2 of this book.

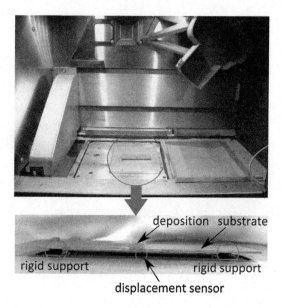

FIGURE 13.2 Experimental setup.

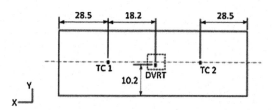

FIGURE 13.3 Bottom view of the workpiece showing the measurement locations (dimensions: mm).

TABLE 13.2 As-used constant material properties and processing conditions

Parameter	Value
Ambient temperature T_∞ [°C]	25 [44]
Argon gas conductivity k_f [W/m/°C]	0.016 [45]
Convection coefficient h [W/m²/°C]	25
Density ρ [kg/m³]	8146 [46]
Laser absorptivity η	0.40 [47]
Poisson's ratio ν	0.30 [48]

The as-used temperature dependent thermal properties for Inconel® 718 are listed in the Appendix. Table 13.2 lists the as-used constant material properties and processing conditions for the analyses.

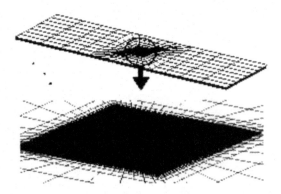

FIGURE 13.4   Finite element mesh for the first layer of deposition.

13.4.2 The Finite Element Mesh

A coarse mesh is created in Patran 2012 (an MSC software) and input into Netfabb Simulation where it is automatically refined. Figure 13.4 shows the mesh used for the first layer of the deposition. The mesh is comprised of 27,963 Hex-8 elements and 46,127 nodes. The mesh is discretized to allow for 1 element per laser radius in the $x-y$ plane of the deposition region and 1 element per deposition thickness in the z direction. The spacial discretization was chosen based on known requirements needed in the Netfabb Simulation solver from the literature [22]. The mesh is coarsened as it moves away from the region that interacts with the heat source. Each node has 1 DOF in the thermal analysis (temperature) and 3 DOF in the mechanical analysis (displacement).

13.4.2.1 Dynamic Mesh Coarsening

The required mesh density for the thermal and mechanical simulations of AM processes is determined by the heat source spot size. The small laser spot size used in LPBF processing makes simulations of the process more computationally expensive than the other AM processes. Due to this fact the task of simulating 91 mm^3 of deposited material using a static mesh would require 3,894,592 Hex-8 elements and 3,510,753 nodes, thus making the approach infeasible. Here a dynamic mesh coarsening scheme is applied on a layer by layer basis. The mesh coarsening strategy allows for elements to be merged below the deposited layer as the heat source advances in the z direction. When elements are merged the interpolation of a variable ϕ is performed as follows:

$$\psi = \frac{1}{8} \sum_{i=1}^{8} \left[\gamma_i \prod_{k=1}^{3} \left(1 + \sqrt{3}\,\xi_k \, \Xi_{ki} \right) \right] \tag{13.1}$$

where γ contains the values of ψ at the Gauss points, ξ are the local coordinates determined by the Newton–Raphson method, and Ξ_{ki} are the local coordinates of the nodes of the element. The coarsening strategy dramatically reduces the number of elements required in the analysis. The mesh used for the final deposition layer requires only 43,904 Hex-8 elements and 61,779 nodes.

13.4.3 Boundary Conditions

The laser heat source Q is modeled using the double ellipsoid model [49]. The convection coefficient h is determined by correlating the thermal results with the experimental temperature measurements and takes into account the conduction losses into the powder outside the deposited region. The mechanical constraints are assigned to match the constraints present in the experiment.

13.4.4 Material Deposition Modeling

The modeling of the deposited material is handled by applying a hybrid quiet/inactive activation strategy developed by Michaleris [41]. The elements belonging to the build structure are initially omitted from the analysis. Here the elements are introduced into the analysis with powder material properties prior to the deposition of their corresponding layer. This simulates the effect of the recoater adding a new layer of metallic powder. The powder thermal properties are applied as calculated in Section 13.2. For the mechanical response simulation the Elastic Modulus of the powder E_{powder} is scaled to reduce its effect on the analysis. It is calculated as:

$$E_{powder} = s_E\, E \qquad (13.2)$$

where $s_E = 10^{-4}$ [22]. The deposition elements are given solid Inconel® 718 properties when they are contacted by the heat source, i.e., when the following condition is met at any one of the eight Gauss points of the element [41]:

$$\frac{6\sqrt{3}}{abc\pi\sqrt{\pi}}\, e^{-[\frac{3x^2}{a^2}+\frac{3y^2}{b^2}+\frac{3(z+vt)^2}{c^2}]} \geq 0.05 \qquad (13.3)$$

13.5 RESULTS AND DISCUSSION

13.5.1 Thermal Results

The thermal model of the deposition process is validated by comparing the calculated temperatures to the thermocouple measurement. Figure 13.5 plots the in-situ temperature results for both the model and the experiment. The simulation overpredicts the maximum measured temperature by 7 °C. Overall, the thermal model captures the trend of the temperature evolution measured during the deposition process.

13.5.2 Mechanical Results

Here, the mechanical simulation results will be viewed along the cross-section of the part shown in Figure 13.6, so that the residual stress in the direction can be more easily observed. Figure 13.7 shows the residual stress distribution calculated after the deposition of the first layer of the part. The tension seen in the deposited layer is due to the cooling and contraction

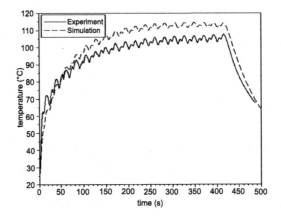

FIGURE 13.5 Comparison of simulated and measured temperature.

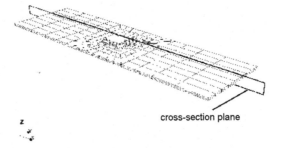

FIGURE 13.6 Cross-section used for viewing mechanical results.

of the molten material after being heated by the laser. The residual stress level reaches 1500 MPa, exceeding the yield strength of the material. The contraction of the molten metal forces the top of the substrate into compression to balance the residual stresses. At this point in the deposition process the x component of the residual stress is the largest due to the fact the deposition passes are performed in the longitudinal direction. This is expected as it is known from the welding literature that the largest residual stresses occur along the longitudinal direction [50].

As additional layers are added, the most recently melted and solidified material will experience the highest levels of tension. Eventually, the lower layers are put into compression by the cooling and contraction of the molten material added above. Figure 13.8 plots the final residual stress distribution after the final layer has cooled, contracted, and plastically deformed. Layers 1 through 20, in addition to the top of the substrate, are in compression. Each layer above layer 20 is in tension. The x and y components of the residual stress are now similar in layers beneath the last deposited layer. This is likely due to the rotating scan pattern, which changes by 67 degrees each layer, causing the residual stress to homogenize.

The residual stress introduced by the moving heat source is the driving force behind the accumulation of distortion in the workpiece. Each time the heat source turns on, the layer being

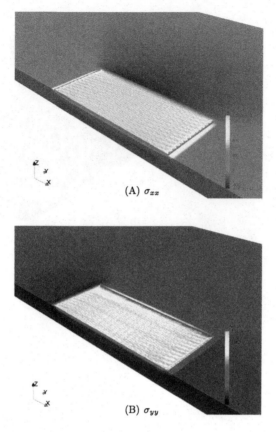

FIGURE 13.7   Residual stress (MPa) distribution after the deposition of layer 1.

melted expands due to thermal effects. The expansion of the top layers causes the substrate to bow upward in the z direction. After the melting cycle is complete, the material is allowed to cool during the recoating period of the process. During the cooling period the molten material contracts, solidifies, and plastifies. The contraction of the molten material causes the substrate to bow downward in the z direction. Due to this, after all of the added material has cooled and contracted, the substrate will have a final shape that is bowed downward in the z direction. This cycle has been observed and described in [51]. Figure 13.9 shows the distortion of the part after the cooling of the final layer.

Experimental in-situ distortion measurements taken using a DVRT were used to thoroughly validate the predicted mechanical response from the model. Figure 13.10 shows the comparison between the predicted and measured distortion results over the course of the entire deposition process. In addition to capturing the overall trend of the distortion accumulation, the model was also successful in calculating the distortion response to the individual layers. The final distortion of the substrate predicted by the model and measured experimentally was 0.826 mm and 0.871 mm, respectively. The model under-predicted the distortion by just 5%.

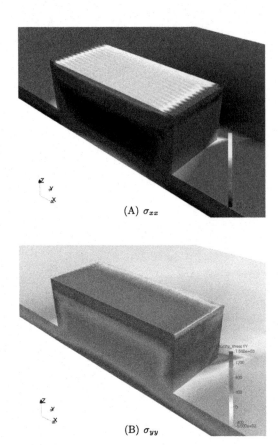

FIGURE 13.8 Residual stress (MPa) distribution after deposition is completed.

FIGURE 13.9 Residual stress (MPa) plotted on the final distorted shape of the workpiece (5× magnification of distortion).

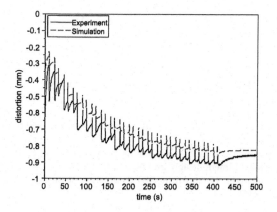

FIGURE 13.10   Comparison of simulated and measured distortion.

13.6 CONCLUSIONS

A finite element modeling strategy was developed to allow for thermoelastoplastic modeling of multi-layer LPBF builds. The modeling strategy takes into account the melting and solidification of each powder layer and assigns the appropriate material properties accordingly. An adaptive meshing strategy was applied to allow for elements to be coarsened on a layer by layer basis, lowering the required DOFs in the system. The conclusions from the study can be summarized as follows:

1. The applied element coarsening strategy significantly reduces the required DOFs present in the analysis. As a result it was possible to perform a thermomechanical simulation for the deposition of 91 mm^3 of material, significantly larger than anything available in the current literature.
2. The model provides insight into how residual stress accumulates in multi-layer builds. Newly deposited layers experience high levels (above the yield strength) of tension. Layers beneath the recently solidified material are forced into tension.
3. The residual stress magnitude observed in the deposited material is greatest in the longitudinal direction of the scan, an observation that agrees with the previously established welding literature.
4. The contraction of the molten metal drives distortion in the workpiece, causing a final distorted workpiece shape exhibiting a downward bow of the substrate.
5. The model achieved strong agreement (5% error) when compared with in-situ distortion measurements taken on an actual deposition of the same geometry.

References

[1] Hibbitt HD, Marcal PV. A numerical, thermo-mechanical model for the welding and subsequent loading of a fabricated structure. Comput Struct 1973;3(5):1145–74.
[2] Friedman E. Thermomechanical analysis of the welding process using the finite element method. J Press Vessel Technol 1975;97:206.

[3] Andersson B. Thermal stresses in a submerged-arc welded joint considering phase transformations. J Eng Mater Technol (United States) 1978;100.

[4] Argyris JH, Szimmat J, Willam KJ. Computational aspects of welding stress analysis. Comput Methods Appl Mech Eng 1982;33(1):635–65.

[5] Papazoglou V, Masubuchi K. Numerical analysis of thermal stresses during welding including phase transformation effects. J Pressure Vessel Technol (United States) 1982;104(3).

[6] Free JA, Porter Goff RF. Predicting residual stresses in multi-pass weldments with the finite element method. Comput Struct 1989;32(2):365–78.

[7] Tekriwal P, Mazumder J. Finite element analysis of three-dimensional transient heat transfer in GMA welding. Weld J 1988;67(5):150–6.

[8] Michaleris P, Tortorelli DA, Vidal CA. Analysis and optimization of weakly coupled thermoelastoplastic systems with applications to weldment design. Int J Numer Methods Eng 1995;38(8):1259–85.

[9] Lindgren LE, Runnemalm H, Näsström MO. Simulation of multipass welding of a thick plate. Int J Numer Methods Eng 1999;44(9):1301–16.

[10] Asadi M, Goldak JA. An integrated computational welding mechanics with direct-search optimization for mitigation of distortion in an aluminum bar using side heating. J Manuf Sci Eng 2014;136(1):011007.

[11] Lindgren LE. Finite element modeling and simulation of welding. Part 1: increased complexity. J Therm Stresses 2001;24(2):141–92.

[12] Lindgren LE. Finite element modeling and simulation of welding. Part 2: improved material modeling. J Therm Stresses 2001;24(3):195–231.

[13] Lindgren LE. Finite element modeling and simulation of welding. Part 3: efficiency and integration. J Therm Stresses 2001;24(4):305–34.

[14] Gouge MF, Heigel JC, Michaleris P, Palmer TA. Modeling forced convection in the thermal simulation of laser cladding processes. Int J Adv Manuf Technol 2015;1–14.

[15] Kolossov S, Boillat E, Glardon R, Fischer P, Locher M. 3D FE simulation for temperature evolution in the selective laser sintering process. Int J Mach Tools Manuf 2004;44(2):117–23.

[16] Peyre P, Aubry P, Fabbro R, Neveu R, Longuet A. Analytical and numerical modelling of the direct metal deposition laser process. J Phys D, Appl Phys 2008;41(2):025403.

[17] Qian L, Mei J, Liang J, Wu X. Influence of position and laser power on thermal history and microstructure of direct laser fabricated Ti-6Al-4V samples. Mater Sci Technol 2005;21(5):597–605.

[18] Shen N, Chou K. Thermal modeling of electron beam additive manufacturing process–powder sintering effects. In: Proceedings of the ASME 2012 international manufacturing science and engineering conference MSEC2012-7253; 2012. p. 1–9.

[19] Jamshidinia M, Kong F, Kovacevic R. Numerical modeling of heat distribution in the electron beam melting® of Ti-6Al-4V. J Manuf Sci Eng 2013;135(6):061010.

[20] Sammons PM, Bristow DA, Landers RG. Height dependent laser metal deposition process modeling. J Manuf Sci Eng 2013;135(5):054501.

[21] Denlinger ER, Irwin J, Michaleris P. Thermomechanical modeling of additive manufacturing large parts. J Manuf Sci Eng 2014;136(6):061007.

[22] Denlinger ER, Heigel JC, Michaleris P. Residual stress and distortion modeling of electron beam direct manufacturing Ti-6Al-4V. Proc Inst Mech Eng, B J Eng Manuf 2014;229:1803–13.

[23] Heigel J, Michaleris P, Palmer T. In situ monitoring and characterization of distortion during laser cladding of Inconel® 625. J Mater Process Technol 2015.

[24] Heigel J, Michaleris P, Reutzel E. Thermo-mechanical model development and validation of directed energy deposition additive manufacturing of Ti-6Al-4V. Addit Manuf 2014.

[25] Anca A, Fachinotti VD, Escobar-Palafox G, Cardona A. Computational modelling of shaped metal deposition. Int J Numer Methods Eng 2011;85(1):84–106.

[26] Chiumenti M, Cervera M, Salmi A, Agelet de Saracibar C, Dialami N, Matsui K. Finite element modeling of multi-pass welding and shaped metal deposition processes. Comput Methods Appl Mech Eng 2010;199(37):2343–59.

[27] Lundbäck A, Lindgren LE. Modelling of metal deposition. Finite Elem Anal Des 2011;47(10):1169–77.

[28] Marimuthu S, Clark D, Allen J, Kamara A, Mativenga P, Li L, et al. Finite element modelling of substrate thermal distortion in direct laser additive manufacture of an aero-engine component. J Mech Eng Sci 2012.

[29] Mughal M, Fawad H, Mufti R. Three-dimensional finite-element modelling of deformation in weld-based rapid prototyping. Proc Inst Mech Eng, Part C J Mech Eng Sci 2006;220(6):875–85.

[30] Grum J, Žnidaršič M. Microstructure, microhardness, and residual stress analysis of laser surface cladding of low-carbon steel. Mater Manuf Process 2004;19(2):243–58.

[31] Ocelík V, Bosgra J, de Hosson JTM. In-situ strain observation in high power laser cladding. Surf Coat Technol 2009;203(20):3189–96.

[32] Paul R, Anand S, Gerner F. Effect of thermal deformation on part errors in metal powder based additive manufacturing processes. J Manuf Sci Eng 2014;136(3):031009.

[33] Neugebauer F, Keller N, Xu H, Kober C, Ploshikhin V. Simulation of selective laser melting using process specific layer based meshing. In: Proc fraunhofer direct digital manufacturing conf (DDMC 2014); 2014.

[34] Roberts I, Wang C, Esterlein R, Stanford M, Mynors D. A three-dimensional finite element analysis of the temperature field during laser melting of metal powders in additive layer manufacturing. Int J Mach Tools Manuf 2009;49(12):916–23.

[35] Li Y, Gu D. Thermal behavior during selective laser melting of commercially pure titanium powder: numerical simulation and experimental study. Addit Manuf 2014;1:99–109.

[36] Zhang D, Cai Q, Liu J, Zhang L, Li R. Select laser melting of W–Ni–Fe powders: simulation and experimental study. Int J Adv Manuf Technol 2010;51(5):649–58.

[37] Zeng K, Pal D, Gong H, Patil N, Stucker B. Comparison of 3DSIM thermal modelling of selective laser melting using new dynamic meshing method to ANSYS. Mater Sci Technol 2015;31(8):945–56.

[38] Matsumoto M, Shiomi M, Osakada K, Abe F. Finite element analysis of single layer forming on metallic powder bed in rapid prototyping by selective laser processing. Int J Mach Tools Manuf 2002;42(1):61–7.

[39] Song X, Xie M, Hofmann F, Illston T, Connolley T, Reinhard C, et al. Residual stresses and microstructure in powder bed direct laser deposition (PB DLD) samples. Int J Mater Forming 2014;1–10.

[40] Dai D, Gu D. Thermal behavior and densification mechanism during selective laser melting of copper matrix composites: simulation and experiments. Mater Des 2014;55:482–91.

[41] Michaleris P. Modeling metal deposition in heat transfer analyses of additive manufacturing processes. Finite Elem Anal Des 2014;86:51–60.

[42] Goldak JA, Akhlaghi M. Computational welding mechanics. Springer Science & Business Media; 2006.

[43] Zhang L, Reutzel E, Michaleris P. Finite element modeling discretization requirements for the laser forming process. Int J Mech Sci 2004;46(4):623–37.

[44] Dunbar A, Denlinger E, Heigel J, Michaleris P, Guerrier P, Martukanitz R, et al. Experimental in situ distortion and temperature measurements during the laser powder bed fusion additive manufacturing process. Part 1: development of experimental method. Addit Manuf 2015. In Review.

[45] Ziebland H, Burton J. The thermal conductivity of nitrogen and argon in the liquid and gaseous states. Br J Appl Phys 1958;9(2):52.

[46] Ahn D, Byun K, Kang M. Thermal characteristics in the cutting of Inconel 718 superalloy using CW Nd: YAG laser. J Mater Sci Technol 2010;26(4):362–6.

[47] Sainte-Catherine C, Jeandin M, Kechemair D, Ricaud JP, Sabatier L. Study of dynamic absorptivity at 10.6 μm (co2) and 1.06 μm (nd-yag) wavelengths as a function of temperature. J Phys IV 1991;1(C7):C7–151.

[48] Yilbas B, Akhtar S, Karatas C. Laser surface treatment of Inconel 718 alloy: thermal stress analysis. Opt Lasers Eng 2010;48(7):740–9.

[49] Goldak J, Chakravarti A, Bibby M. A new finite element model for welding heat sources. Metall Trans B 1984;15(2):299–305.

[50] Deo M, Michaleris P. Mitigation of welding induced buckling distortion using transient thermal tensioning. Sci Technol Weld Join 2003;8(1):49–54.

[51] Denlinger ER, Heigel JC, Michaleris P, Palmer T. Effect of inter-layer dwell time on distortion and residual stress in additive manufacturing of titanium and nickel alloys. J Mater Process Technol 2015;215:123–31.

Study of the Evolution of Distortion During the Powder Bed Fusion Build Process Using a Combined Experimental and Modeling Approach*

Alexander J. Dunbar

Applied Research Laboratory, The Pennsylvania State University, State College, PA, USA

14.1 INTRODUCTION

Recent developments in additive manufacturing (AM) allow for the rapid production of end-use parts without the need for significant post-build machining. In particular, the accuracy attainable by laser powder bed fusion (LPBF) AM allows for parts to be built with complex interior geometries previously unattainable by traditional manufacturing means. However, residual stresses caused by localized thermal gradients in the parts often result in build failures. Build failures include: delamination of layers, support structure fracture, part interference with the recoating mechanism, high levels of post-build distortion, voids in solid material, and lack of fusion of deposited material [1–3]. In order to mitigate the several types of failure associated with high levels of material deformation, an improved understanding of how deformation and residual stress accumulate in AM parts is needed.

Much of the experimental research is focused on the micro-scale, single lines and individual layers. Pohl et al. [4] varied laser scan pattern and speed for a single layer and used post-process experimental deflection results to compare and identify the best build parameters. Work shown by Kempen et al. [5] utilizes a new technique for in-situ melt pool characterization. By analyzing the melt pool in-situ, ideal processing parameters are selected

---

* Reprinted from Additive Manufacturing, Vol 12, Dunbar, Alexander J and Denlinger, Erik R and Gouge, Michael F and Michaleris, Pan, Experimental validation of finite element modeling for laser powder bed fusion deformation, Pages No. 108–120, Copyright (2016), with permission from Elsevier.

by performing several single track tests. While experiments like these are useful in the determination of ideal processing parameters required to achieve fully dense and fused material, they do not address geometry dependent structural stresses that build up in full-scale LPBF built components.

While experiments may be informative to the LPBF build process, when extended to the larger scale this type of parametric analysis proves to be expensive. In these cases, simulation can be used to identify ideal parameters to be used in LPBF. Zhang et al. [6] performed an analysis on processing parameters for the LPBF build process. The study focused on a W-Ni-Fe powder bed build where a model was used to determine processing parameters required to reach a user specified melt pool depth. Another study performed by Dai et al. [7] used simulations to predict melt pool size based on an input laser energy density. The work focused heavily on measuring the size of the melt pool and determined that the melt pool dimensions are on the order of hundreds of microns for LPBF. These studies provide examples of how modeling can be used in the setup stages of LPBF to identify processing parameters for a specific build process without the need for expensive experiments.

Modeling of additive manufacturing has proven to be successful in improving the understanding of the distortion accumulation during the AM process and guiding design to reduce build failures [8–11,5,12–21]. For example, Denlinger et al. [22] demonstrates the utility that FE modeling can have in the prevention of part distortion. Similar models are scarce in the powder bed field. Most models that are available are limited to simulating small build volumes or single layers [17,16], melt pool dynamics [7] or have limited their analysis to the thermal side of the LPBF process [16,6,14]. While each of the referenced studies provide utility to the field of additive manufacturing, none has demonstrated capabilities for full part-scale modeling.

The aforementioned models are limited in implementation with powder bed AM systems in that they are too computationally expensive for application in full part-scale modeling of LPBF. Mesh refinement studies show that element size must be at least the size of the melt pool radius or temperature equations will not properly converge [8]. For general purpose FE software (e.g., ANSYS, Abaqus) at least four elements are required per heat source radius [16]. Therefore, models that are applicable to alternate AM processes (e.g., Directed Energy Deposition) cannot be directly applied to LPBF processes, as the heat source diameter is approximately 10 times smaller for LPBF as compared to Directed Energy Deposition. This reduction in heat source diameter results in a minimum 1000 fold increase in elements to capture a similar volume of material. Currently, there are two known available software packages that are capable of modeling part-scale LPBF: (1) Diablo [11] and (2) Netfabb Simulation. To accurately model part-scale LPBF, special consideration must be made to include the complex inter-layer effects, resulting from the complex thermal history of the parts [23]. For that reason, Netfabb Simulation is used, as it allows for accurate modeling of inter-layer effects.

Many studies focus on the micro-scale (single tracks or layers) of the LPBF process. While the smaller scale studies are necessary in defining the mechanics of LPBF, understanding how a larger part distorts as a result of the build process is an important step in build failure mitigation for LPBF. In contrast to the abundance of studies focusing on the melt pool, process parameters and microstructure, few have published final build geometry of LPBF-made parts. Here, two experimental builds of a simple cylindrical geometry with differing

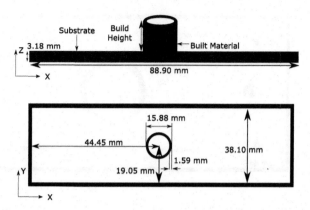

FIGURE 14.1 Schematic of the deposition. Note that the build height is not constant for both cases. Build height for Case 1 (rotating scan pattern) is 6.16 mm. Build height for Case 2 (constant scan pattern) is 12.70 mm.

scan patterns are deposited to measure final build geometry and distortion profiles for parts made with LPBF AM. In addition to experimental analysis, a finite element (FE) model is experimentally validated and applied. The experiment performed also provides necessary validation data for future LPBF models. The use of modeling software for this work is to help further understanding of stress and distortion evolution throughout the build process.

14.2 EXPERIMENT

14.2.1 Experimental Setup

Two LPBF builds are used to measure the distortion profiles of parts made using the LPBF build process. Experiments are designed with similar cross-sectional geometry with the goal of studying the effects of scan pattern orientation. Parts are designed to be modeled allowing further analysis on the evolution of stress and distortion. By reducing the complexity of the part, analysis is made easier as part geometries are constant along the height of the part.

14.2.1.1 Description of Experimental Builds

Build dimensions and locations are defined by user created computer aided design (CAD) files. Builds are completed using interchangeable substrates. Interchangeable substrates allow builds to be removed from the LPBF machine without requiring post-build machining which will affect final build geometries. Figure 14.1 is a schematic of build geometry and substrate for Case 1 and Case 2. The substrate dimensions are 88.9 mm × 38.1 mm × 3.18 mm in each case. Both cases are cylindrical geometries with an outer diameter of 15.88 mm and a cylinder wall thickness of 1.59 mm. As a result of a machine failure during production, build heights are 6.16 mm and 12.70 mm for Case 1 and Case 2, respectively. While the difference in part height prevents direct comparisons between the two cases, each case can still be analyzed as the build progressed properly until failure. Substrates are attached to the *vault*

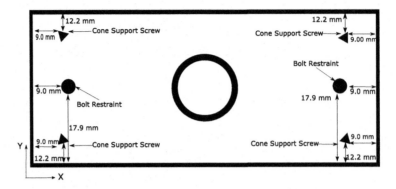

FIGURE 14.2 Schematic of the constraints used to hold the substrates during the build process.

measurement system [23]. Substrates are constrained from underneath using restraining bolts that are threaded into holes on substrates, thereby restricting motion of the substrate without impacting the recoater mechanism. Substrates are supported in the corner using cone head set screws. Dimensions for constraints are shown in Figure 14.2.

Figure 14.3 shows the fully built material for Cases 1 and 2 with their substrates. In addition to differences in part height, Case 1 is built with the machine default rotating laser scan pattern and Case 2 is built with a constant scan pattern parallel with the Y axis. The rotating versus constant scan pattern is used to determine how alteration of this build parameter affects final build geometry. A schematic demonstrating a rotating versus constant scan pattern is shown in Figure 14.4. Descriptions of Cases 1 and 2 processing parameters are summarized in Table 14.1.

The substrate and powder materials are both Inconel® 718. The average powder diameter is 30.4 ± 7 μm. Inconel® 718 is used as it is a common superalloy used in AM across several industries [24–28]. Both cases are built using the EOS M280 LPBF machine. The EOSINT M280 machine uses a 4LR-400-SM-EOS laser that operates at a wavelength of 1060–1100 nm with a power of 280 W. Processing parameters used for these experiments include a layer thickness of 40 μm, a hatch spacing of 110 μm, a laser travel speed of 960 mm/s, and a laser power of 280 W.

14.2.1.2 Description of Measurement Equipment

Post-process measurements of the final build geometry are performed to quantify final distortion in each of the cases. A Core® RS-50 coordinate-measuring machine (CMM) is used to measure the outside cylinder walls at 4 separate XY positions for Cases 1 and Case 2. The CMM uses a positive contact probe to measure the distortion profile along the height (Z) of the part. Measurement locations are shown in the schematic in Figure 14.5. Measurement locations are labeled as +X, +Y, −X, and −Y, which correspond to the X and Y axis with an origin at the center of the build, these labels will be used when comparing and contrasting results. Measurements are made approximately every 0.05 mm along the height of the part. Measurements at the base of the part, lower than approximately $z = 1.5$ mm, were not captured due to the size of the measurement probe used. A recent calibration, completed in accordance ASME B894.1, Sections 5.3, 5.4.3, 5.5.2, and 5.5.4, shows a measurement accuracy

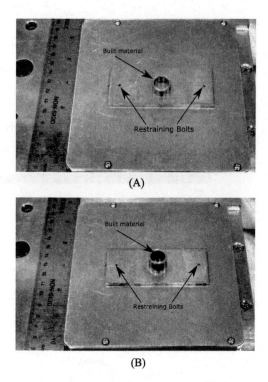

FIGURE 14.3 Completed build with substrate for: (A) Case 1 (rotating scan pattern); (B) Case 2 (constant scan pattern).

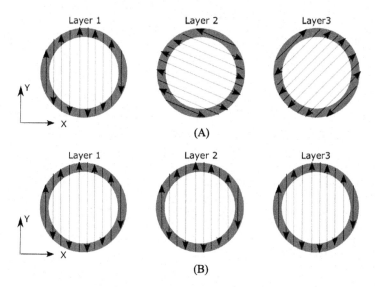

FIGURE 14.4 Schematic of the scan pattern for: (A) Case 1 (Rotating Scan Pattern); (B) Case 2 (Constant Scan Pattern).

TABLE 14.1   Description of experimental cases

Case Number	1	2
Material	Inconel®718	Inconel®718
Layer Thickness μm	40	40
Laser Speed (mm/s)	960	960
Hatch Spacing (mm)	0.11	0.11
Rotating Scan Pattern	Yes	No

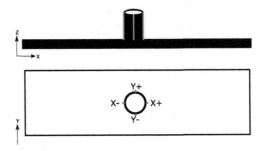

FIGURE 14.5   Schematic of CMM measurement locations for Cases 1 and 2.

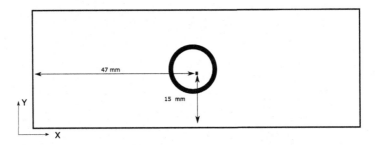

FIGURE 14.6   Measurement location for in-situ distortion measurements in the build direction (Z) during the build process.

of ±0.0044 mm over 1200 mm, ±0.0032 mm over 1000 mm, and ±0.0020 mm over 750 mm in the x, y, and z direction respectively. Distortion of the substrate during the build process is captured using a Lord Microstrain M-DVRT-3 connected to a DEMOD-DC 2 signal conditioner which translates displacements into voltages which are recorded using a National Instruments USB-6009 data acquisition (DAQ) system. This system provides in-situ measurements of substrate distortion in the Z direction with an accuracy of ±15 μm. The measurement location for the Z distortion is found in Figure 14.6.

14.2.2 Experimental Results

Figure 14.7 shows the experimental measurements for both Case 1 and Case 2 along the four measurement locations (Figure 14.5). In this study, distortion is defined as the difference between the prescribed and measured positions at the measurement locations. For both

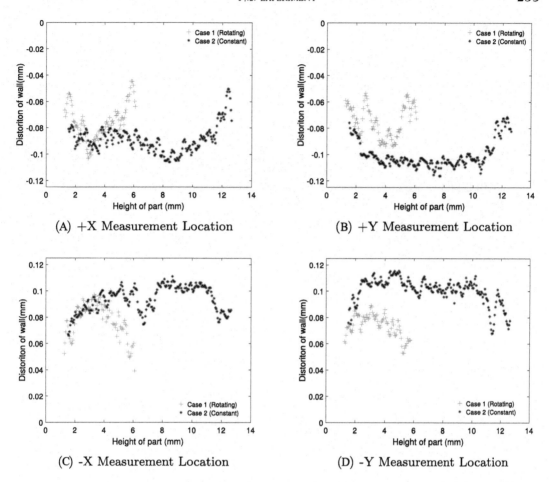

FIGURE 14.7 Distortion measurement results along the four measurement locations for both Case 1 (rotating scan pattern) and Case 2 (constant scan pattern).

cases, oscillation in the distortion measurements is seen along the height of the part. Given the previously stated accuracy of the CMM machine, this is likely an effect of high surface roughness, which is common for metal powder based AM [29,30].

The distortion measurement for Case 1 at the positive X location distortion is nearly parabolic, with magnitude of distortion reaching a peak (0.10 mm) at only a single height (2.86 mm), whereas the distortion profile for Case 2 (constant scan pattern) has a nearly constant magnitude of distortion (0.92 mm) along much of the part height (from approximately 2 mm to 10 mm) when excluding oscillation due to surface roughness. Figure 14.7B shows measurements at the positive Y measurement location. Consistent distortion shape profiles are seen for both Case 1 (rotating scan pattern) and Case 2 (constant scan pattern), with surface roughness affecting the measurements for Case 1 (rotating scan pattern) between a part height of 2 mm and 4 mm. Measurements of the negative X location are shown in

TABLE 14.2   Case 1 (rotating scan pattern) peak distortion comparison by percent deviation with experimental measurements

Part Height (mm)	Mean Distortion Magnitude (mm)	Positive X Position (%)	Positive Y Position (%)	Negative X Position (%)	Negative Y Position (%)
Z = 2.0	0.079	1.3	2.3	1.6	2.6
Z = 3.0	0.086	13.6	14.8	4.3	3.1
Z = 4.0	0.083	1.5	6.3	1.6	6.5
Z = 5.0	0.072	1.8	4.3	0.8	6.9
Z = 5.5	0.063	0.6	0.7	5.2	5.2

TABLE 14.3   Case 2 (constant scan pattern) peak distortion comparison by percent deviation with experimental measurements

Part Height (mm)	Mean Distortion Magnitude (mm)	Positive X Position (%)	Positive Y Position (%)	Negative X Position (%)	Negative Y Position (%)
Z = 2.0	0.087	1.7	1.6	7.5	7.3
Z = 4.0	0.098	12.7	7.8	3.3	8.1
Z = 6.0	0.099	10.7	8.8	0.4	1.3
Z = 8.0	0.103	1.4	2.3	2.2	3.2
Z = 10.0	0.102	6.8	1.5	1.3	4.0
Z = 12.0	0.082	13.3	1.8	1.0	14.1

Figure 14.7C. Similar shape profiles are again present for both Case 1 (rotating scan pattern) and Case 2 (constant scan pattern) as with Figure 14.7A and Figure 14.7B. For both X measurement locations (Figure 14.7A and Figure 14.7C) indentations in distortion profile can be seen between build heights of 4–8 mm, this is not seen on either of the Y measurement locations (Figure 14.7B and Figure 14.7D). Measurements of the −Y measurement location, shown in Figure 14.7D, compare well with the other measurement locations with similar magnitude and shape for both cases. For both cases and each measurement location, the part distorts inward toward the center line of the cylindrical geometry.

Distortion values were compared for each of the four measurement locations at several heights along the part for both cases. At each height, a mean value of distortion is calculated, which is the averaged magnitude of the measured distortion between the four measurement locations. For those same heights, a percent deviation from this mean is calculated for each measurement location, to identify any trends in distortion profiles that may align with either the X or Y directions. These measurements are located in Table 14.2 and Table 14.3 for Case 1 (rotating scan pattern) and Case 2 (constant scan pattern) respectively. To account for surface roughness, measurements at specified heights are averaged with the five measurement below and above selected height.

Direct comparison cannot be made between experimental results of Case 1 and Case 2, as a result of their differing build heights but, within each case the four measurement locations can still be compared to each other. Tables 14.2 and 14.3 show that for both Cases 1 and 2 distortion measurements are consistent (under 15% difference) for each direction with no increased distortion trend aligned with either the X or Y directions. For Case 1, a rotating scan pattern is used, so isotropic distortion profiles are expected for each measurement location. For Case 2, all laser scans are aligned with the Y axis (Figure 14.4) so an asymmetry is expected for the distortion profiles. In particular, it is expected that the Y (longitudinal) distortion magnitudes would be larger than the X (transverse) distortion profiles [3]; however, no trend is found in the distortion profiles for Case 2 (constant scan pattern). Instead, large deviations found in any measurement location seem to be the result of localized defects, identified by large deviations from the mean distortion profile at only a single height with no trend continuing along the entire part height.

For each measurement location, distortion profiles for Case 2 (constant scan pattern) plateau at a part height of 3 mm, whereas the distortion profile for Case 1 (rotating scan pattern) reaches a peak distortion at one height resulting in a shape that is most nearly parabolic and where no plateau is reached. This is a result of their relative build heights where Case 1 is 6.16 mm tall and Case 2 is approximately 12.70 mm tall. For the geometry used in these experiments, at a height of 4 mm, the rate at which distortion increases with height levels off until near the top of the part at approximately 10 mm. At low build heights the deposited material is constrained by the substrate thereby reducing the amount of distortion possible. As the part increases in height during the build process, the constraining effects of the substrate are minimized, and the part can distort freely. Distortion for these geometries is caused by the melted material of the current layer solidifying and compressing the material below it. For both cases, material near the top of the part distortion decreases rapidly as there are not enough layers above this point to cause significant distortion. To further examine these effects, an FE model is used to provide analysis and to help understand the distortion evolution throughout the build process.

14.3 POWDER BED FUSION SIMULATION

The thermal and mechanical histories are determined by performing a three-dimensional transient thermal analysis and a three-dimensional quasi-static incremental analysis, respectively. The thermal and mechanical analyses are performed independently and are weakly coupled, meaning that the mechanical response has no effect on the thermal history of the workpiece [31].

14.3.1 Numerical Implementation

The thermal and mechanical analyses are performed using the method described in Chapter 2. The analyses are done in a series of time steps with the current time step taking the solution at the previous time step as the initial condition. At each time step, the discrete equilibrium equations are solved by using the Newton–Raphson method.

The temperature dependent thermal properties and mechanical properties for Inconel® 718 can be found in the Appendix. Table 14.4 lists the as-used constant material properties and processing conditions for the analyses.

TABLE 14.4 As-used constant material properties and processing conditions

Parameter	Value
Ambient temperature T_∞ [°C]	25 [23]
Convection coefficient h [W/m^2/°C]	8
Density ρ [kg/m^3]	8146 [28]
Laser absorptivity η	0.40 [32]
Poisson's ratio ν	0.30 [24]

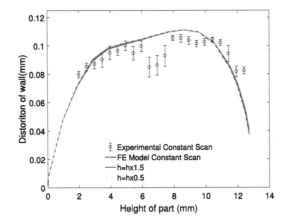

FIGURE 14.8 Sensitivity of post-build distortion to FE model convection coefficient for constant scan pattern case (Case 2).

Loose powder that goes unmelted during the deposition process is not included as part of the analysis. Conduction into the powder is modeled by applying an artificial convective boundary condition on the model. The value for the convection coefficient applied on all surfaces is shown in Table 14.4. Several simulations using varying convection coefficients h were run in order to determine the sensitivity of the model to this unknown parameter. The result of these studies show that for these build geometries, a 50% increase or decrease in the value of convection coefficient result in less than 1% error on final distortion results shown in Figure 14.8. The laser heat source Q is modeled using the double ellipsoid model described in Chapter 2.

Figure 14.9 shows the mesh used for the analysis. Meshing through Netfabb Simulation is done automatically with adaptive meshing allowing for coarse elements sufficiently far from the current layer and reappropriate refinement where accuracy is necessary. Three different mesh densities were simulated for each case to determine if the simulation results were mesh independent. A mesh refinement study was completed for distortion results. For both cases the mesh used for the analysis demonstrated monotonic convergence with a peak difference of distortion magnitude under 2%.

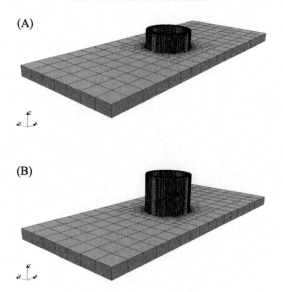

FIGURE 14.9 Mesh used for: (A) Case 1 (rotating scan pattern) simulation; (B) Case 2 (constant scan pattern) simulation.

14.4 RESULTS AND DISCUSSION

14.4.1 Model Comparison to Experimental Measurements

Simulation results were extracted at the experimental measurement locations (Figure 14.5). For ease of comparison and to reduce noise from surface roughness, experimental measurements were averaged in groups of 10 with the standard deviation indicated by error bars. For these simulations, fixed substrate is used for mechanical boundary conditions. These boundary conditions prevent translation or distortion of the substrate. For the thermal analysis, initial conditions set the entire substrate to the ambient temperate ($T_\infty = 25\,°C$).

A comparison between simulation results and experimental measurements for both Case 1 (rotating scan pattern) and Case 2 (constant scan pattern) can be found in Figure 14.10A for the positive X location. Similar comparison for the experimental measurements and simulation results for the positive Y measurement location can be found in Figure 14.10B. Surface roughness, which cannot be captured by the model, affects the comparison between the model and experiments for Case 1 in Figure 14.7B most notably at a height of 3 mm. Simulation results compared with experimental measurements are shown in Figure 14.10C for the negative X location. A significant part defect, caused by earlier described surface effects common to AM, for Case 2 at a part height of 7 mm causes a localized discrepancy, but model results match experimental results outside of this region. The comparison for Case 1 is within measurement averaged, standard deviation indicated by the error bars. The comparison of the simulation results and experimental measurements for the negative Y location is shown in Figure 14.10D. For each case, the

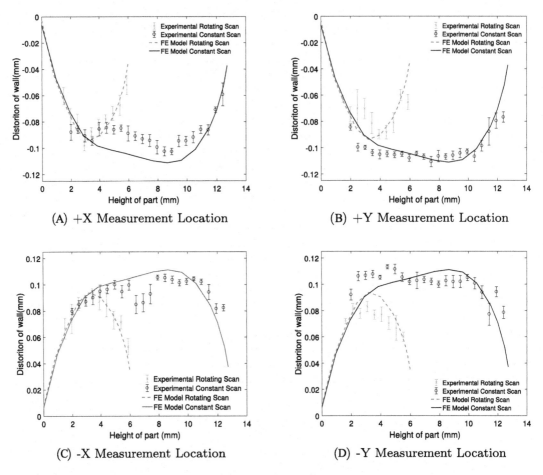

FIGURE 14.10 Averaged CMM measurements of post-build distortion compared against FE model results at the four measurement locations shown in Figure 14.5.

FE model results match the experimental measurements well in both trend and magnitude.

For quantitative comparison between experiment and simulation, comparisons were made along the height of the part in the FE model. Simulation results were averaged at each nodal height along the height of the part for each measurement locations (e.g. $+X$, $+Y$, $-X$, $-Y$). Comparisons for Case 1 and Case 2 are shown in Table 14.5 and Table 14.6, respectively. On average, FE model results compare well with experimental measurements with the largest difference between simulation results and experimental measurements as 12% at the $+X$ measurement location for Case 2. Measurement locations with the largest percent errors correlate with previously discussed defects and roughness in experimental build surfaces.

TABLE 14.5 Comparison of experimental measurements and simulation results for Case 1 (Rotating scan pattern). The error is averaged at each nodal location in the FE model along the height of the part for each measurement location (Figure 14.5). Measurements are normalized by the experimental distortion measurement at each height

Measurement Location	Averaged Error in FE Model Along the Height of the Part (%)
Positive X Position	6.2
Positive Y Position	10.2
Negative X Position	5.1
Negative Y Position	9.0

TABLE 14.6 Comparison of experimental measurements and simulation results for Case 2 (Constant scan pattern). The error is averaged at each node in the FE model along the height of the part for each measurement location (Figure 14.5). Measurements are normalized by experimental distortion measurement at each height

Measurement Location	Averaged Error in FE Model Along the Height of the Part (%)
Positive X Position	11.8
Positive Y Position	4.5
Negative X Position	7.6
Negative Y Position	7.6

14.4.2 Substrate Deformation

Reviewing distortion measurements for Case 2 (constant scan pattern) for the positive and negative X measurement locations, an indentation in the distortion profile can be seen between a build height of approximately 4 mm and 8 mm. Simulations, shown in Figure 14.13 and Figure 14.10, modeled with a rigid substrate are unable to capture this feature in the distortion profile. To further investigate, a simulation with a flexible substrate that more closely matches the boundary conditions defined in Figure 14.2 was completed. To match these constraints, translation in the Z direction is fixed at the cone set screw locations shown in Figure 14.2. In addition, the X, Y and Z translations were fixed at the threaded bolt locations shown in Figure 14.2. FE model results of the distortion of the substrate in the build direction (Z) are compared against in-situ experimental measurements. A comparison of simulation results and experimental measurements of the distortion in the build direction is found in Figure 14.11. At approximately 2400 seconds, the measurement equipment lost power and stopped recording, this is identified in Figure 14.11 by a vertical dotted line. The model result for Z distortion is within 0.01 mm placing it within 10% of the experimental measurement.

Results for each of the measurement directions are shown in Figure 14.12. From the results shown in Figure 14.12, it can be surmised that the localized indentation in the X measurement locations are likely caused by the flexible substrate. This effect is likely driven by the X measurement locations being aligned with the long dimension of the substrate. As the substrate distorts in the negative Z direction, it causes an indentation in the cylinder wall. This

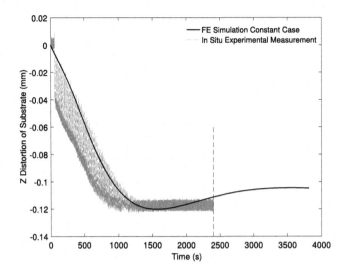

FIGURE 14.11 Comparison of experimental and FE model results for distortion of the substrate in the build direction.

TABLE 14.7 Comparison of experimental measurements and simulation results for Case 2 (Constant scan pattern) with and without a flexible substrate. The error is averaged at each node in the FE model along the height of the part for each measurement location (Figure 14.5). Measurements are normalized by experimental distortion measurement at each height

Measurement Location	Averaged Error in FE Model Along the Height of the Part Fixed Substrate (%)	Averaged Error in FE Model Along the Height of the Part Flexible Substrate (%)
Positive X Position	11.8	8.6
Positive Y Position	4.5	7.8
Negative X Position	7.6	5.2
Negative Y Position	7.6	9.4

result is concluded by comparing the X distortion profiles (shown in Figure 14.12A and Figure 14.12C) and the Y distortion profiles (shown in Figure 14.12B and Figure 14.12D). For the Y measurement locations, the inclusion of the flexible substrate does not change the shape of the distortion profile, only the magnitude. However, for both of the X measurement locations, a distinct change in the distortion profile can be seen, which more closely matches the experimental distortion profile for the X direction. As no indentation is observed for Case 1 (rotating scan pattern), it is also likely that the indentation is also affected by part height. Inclusion of the flexible substrate in the simulation improves the accuracy as compared with experimental measurements. Table 14.7 presents quantitative results demonstrating improved accuracy of the flexible substrate simulation, using the previously described method for determining model accuracy to experimental measurements. Results shown in Table 14.7 show increased simulation accuracy in the positive X and negative X directions, but at a cost of accuracy in the Y directions.

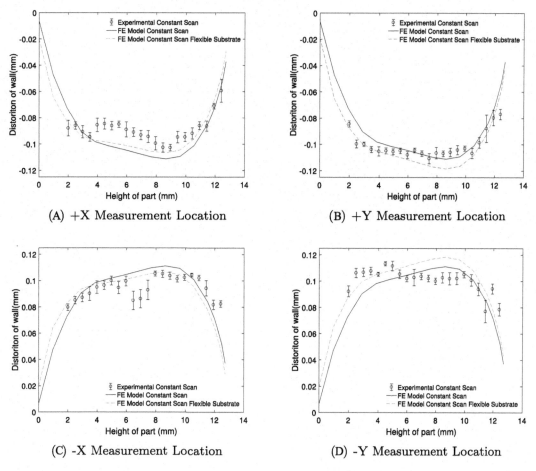

(A) +X Measurement Location

(B) +Y Measurement Location

(C) -X Measurement Location

(D) -Y Measurement Location

FIGURE 14.12 Distortion results for simulation of Case 2 (Constant Scan Pattern) with and without a flexible substrate for each of the four measurement locations shown in Figure 14.5.

14.4.3 Extension of Rotating Scan Pattern Case

Due to a shortage of powder during the build process, the rotating scan pattern case was built shorter than designed. Using the previously validated model, the rotating scan pattern model was extended to its prescribed build height. The results of the extended rotating scan pattern case are shown in Figure 14.13.

Results from these simulations (shown in Figure 14.13) conform previous statements regarding the minimal effect of rotating scan pattern for the thin-walled geometry presented. For each of the measurement locations, the rotating scan pattern consistently distorts less than the constant scan pattern, but the difference between the two cases are on average 1–2%, well within the models determined accuracy. The use of the FE model to provide a comparison of a part that was not successfully built with one that built correctly highlights another potential applications for FE model usage in LPBF AM.

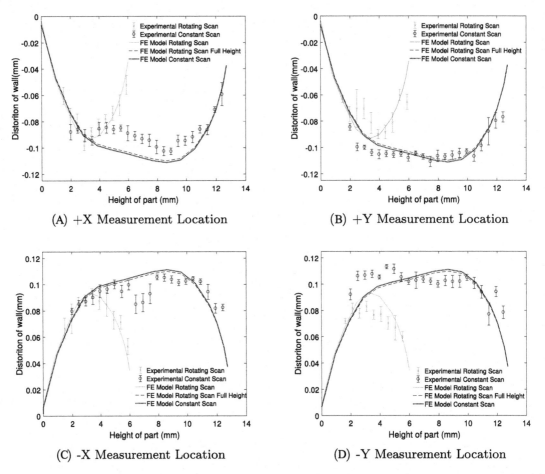

FIGURE 14.13 Extension of the rotating scan pattern case to full build size for the four measurement locations shown in Figure 14.5.

14.4.4 Distortion Evolution in Time

For both cases, the final shape of the deposited cylinder flares out at the bottom where attached to the substrate, and the top where uncompressed by subsequent layers.

Figure 14.14 shows the simulation prediction of the build geometry, with distortion magnified 10 times, at the halfway (Figure 14.14A and C and final Figure 14.14B and D) build time for Cases 1 and 2. The distortion magnitude for the constant scan pattern case (Case 2) indicates that at halfway through build, shown in Figure 14.14C, the part has already reached the peak distortion magnitude, whereas for the rotating scan pattern, maximum distortion at halfway through the build (Figure 14.14A) is lower than the final distortion values. The change in the profile of the part from Figure 14.14A to 14.14B and Figure 14.14C to 14.14D shows that the maximum distortion is not located at the top layer; rather, it is located at a height several layers beneath the top layer. This strengthens the conclusion that for this

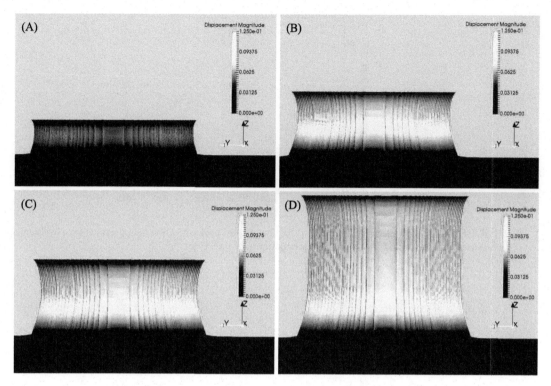

FIGURE 14.14 Contour plots of distortion in mm (distortion magnified 10×) for: (A) Case 1 (rotating scan pattern) halfway through build; (B) Case 1 (rotating scan pattern) finished build after part has cooled to ambient temperature; (C) Case 2 (constant scan pattern) halfway through build; (D) Case 2 (constant scan pattern) finished build after part has cooled to ambient temperature.

geometry, distortion is caused by the multiple layers solidifying and compressing layers previously built. The accumulation of tensile stresses induced by the cooling and contracting of newly deposited material forces the middle height portion of the part into compression. For all four contour plots, distortion is a local minimum at the top layer whether halfway through the build or the finished build.

Figure 14.15 compares the distortion value of the current top layer and the maximum distortion versus time for Case 2. At time $t = 4000$ s, the build process is completed, and the simulation allows the part to cool back to ambient temperature ($T = 25\,°C$), causing the part to contract, resulting in an increase of the magnitude of the distortion profile. The importance of Figure 14.15 is to demonstrate that from extremely early moment in the build (approximately $t = 100$ s) peak distortion is not at the top layer, but instead at a lower layer. Therefore, distortion in these geometries is not a single layer effect, but instead it is the cumulative effect of multiple layers.

Figure 14.16 shows the current height of the part and the height along the wall where the peak distortion is located for Case 2. Although, the location of the peak distortion continues to rise throughout the build process, it climbs slower than the height of the part.

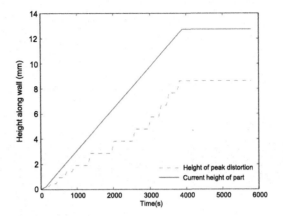

FIGURE 14.15 Comparison of the current part height and peak distortion height versus time for the negative X measurement location for Case 2 (constant scan pattern).

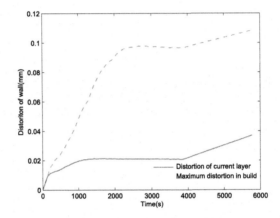

FIGURE 14.16 Comparison of the distortion of the top layer and the peak distortion value versus time for the negative X measurement location for Case 2 (constant scan pattern).

For the majority of the build process, peak distortion is not found at the top layer, but instead at the layers below. During the build process, initially the peak distortion is located at the current layer, but once a sufficient part height is reached approximately (6.5 mm), the part height where the current peak distortion is located lags behind the height of the current layer. This can be seen in Figure 14.16. Additionally, once the build progresses to approximately 2000 s (build height of 6.24 mm), the value of the peak distortion plateaus. The plateau of distortion shown in Figure 14.16 shows that once a sufficient build height is reached, peak distortion will become independent of part height. This effect determines the final shape of cylinders made in a LPBF build process which are flared out at the bottom and top of the part as the part restricted by the substrate below it and the top is relatively undistorted as there is no material above it to compress it. Results from the model give insight to distortion accumulation process and how currently added layers can affect previous layers.

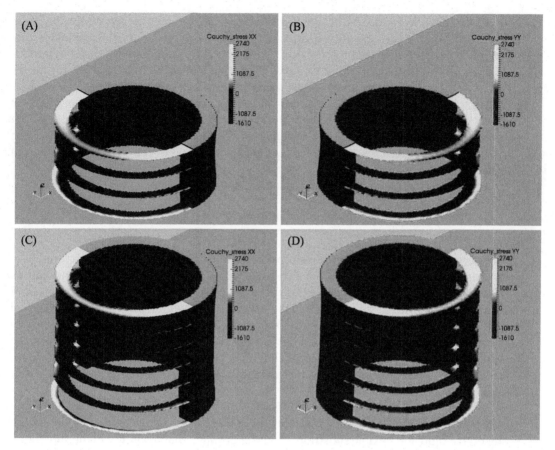

FIGURE 14.17 Case 2 (constant scan pattern) contour plots of Cauchy stress in MPa (distortion magnified 2.5×) for: (A) Cauchy XX stress halfway through build (Z = 6.3 mm); (B) Cauchy YY stress halfway through build (Z = 6.3 mm)' (C) Cauchy XX stress for finished build; (D) Cauchy YY stress for finished build.

Examining stress contours calculated by the model can help to explain the final shape of the part. X and Y principle Cauchy stresses are shown in Figure 14.17 at various heights through the build at halfway through the build process and for the completed part. Figure 14.17 shows the bottom and top layers are in tension with high stress magnitudes. For each of the measurement locations, the distortion is minimal at the top and bottom of the part. In the middle of the part compressive stresses dominate, reaching a maximum (approximately −1600 MPa) between a part height of 2 mm and 10 mm. Distortion reaches its peak magnitude in this region. Cauchy XX and YY stress profiles shown in Figure 14.17C and Figure 14.17D, demonstrate this transition from tension at the bottom to compression throughout much of the part height back to tension at the top of the part. The result of this is a part that is undistorted at the top and bottom and with significant distortion between those section, resulting in an hourglass-like shape.

As is shown in Figure 14.14, the built material final geometry resembles the previously described hourglass shape with distortion toward the centerline of the cylinder's geometry for the majority of the height of the part. When comparing Cauchy XX Stress (Figure 14.17A with Figure 14.17C) and Cauchy YY stress (Figure 14.17B with Figure 14.17D) for the halfway completed and fully completed builds, the maximum and minimum stress values at halfway through the build are within 5% of the completed build minimum and maximum stresses. Stress calculations are consistent with previous statements that the magnitude of distortion is independent of height, once a sufficient part height is reached for this geometry. For these parts, distortion and stress are caused by the accumulation of compression effects from multiple layers.

14.5 CONCLUSION

Two experimental builds with a simple cylindrical geometry are manufactured in a LPBF machine, one built with a rotating scan pattern and the other with a constant scan pattern, with the goal of providing measurements of the post-build distortion. CMM measurements of distortion show that for both cases, the parts distort towards the center line of the cylinder with approximately equal magnitude for each measurement location for both cases. Experimental measurements are compared against the Netfabb Simulation FE model for validation purposes. FE model results compare well with experimental measurements made. The highest averaged percent error for any measurement location distortion profile falls within 12% of the experimental measurement.

Netfabb Simulation has been demonstrated to effectively model in-situ distortion of the substrate in the direction of the build height (Z) within 10%. By completing simulations using both a rigid and flexible substrate, it was determined that flexibility in the substrate affects the shape of the distortion profile for the X direction. The inclusion of the flexible substrate for these simulations increases accuracy in the X measurement locations by 2–3% albeit at the cost of accuracy in the Y directions. Additional utility of the model is also demonstrated in extension of the rotating scan pattern case (Case 1), which was under-built due to complication during the build process. The FE model first validated on the experimental build of Case 1, extended the build height of that case to the predefined build height for comparison with the constant scan pattern case (Case 2). Comparison of the model results for these cases show little discrepancy, further demonstrating that a rotating versus constant scan pattern does not have a significant impact on these thin walled geometries.

The FE model used for these simulation provides new information regarding the evolution of distortion during the build process. For these thin-walled cylindrical geometries, distortion and residual stress accumulation is caused by solidifying material in the current layer compressing previous layers causing the part to distort inwards. As a result, distortion in the current top layer is typically small (less than 30% of the peak distortion) during the build process and peak distortion is typically several layers below the top layer. Stress profiles calculated using the FE model are used to further explain the distortion profile along the height of the part.

References

[1] Kruth J, Froyen L, Van Vaerenbergh J, Mercelis P, Rombouts M, Lauwers B. Selective laser melting of iron-based powder. J Mater Process Technol 2004;149(1):616–22.

[2] Yadroitsev I, Bertrand P, Smurov I. Parametric analysis of the selective laser melting process. Appl Surf Sci 2007;253(19):8064–9.

[3] Deo M, Michaleris P. Mitigation of welding induced buckling distortion using transient thermal tensioning. Sci Technol Weld Join 2003;8(1):49–54.

[4] Pohl H, Simchi A, Issa M, Dias H. Thermal stresses in direct metal laser sintering. In: Proceedings of the 12th solid freeform fabrication symposium; 2001.

[5] Kempen K, Thijs L, Van Humbeeck J, Kruth J. Processing AlSi10Mg by selective laser melting: parameter optimisation and material characterisation. Mater Sci Technol 2014;31(8):917–23.

[6] Zhang D, Cai Q, Liu J, Zhang L, Li R. Select laser melting of W–Ni–Fe powders: simulation and experimental study. Int J Adv Manuf Technol 2010;51(5):649–58.

[7] Dai D, Gu D. Thermal behavior and densification mechanism during selective laser melting of copper matrix composites: simulation and experiments. Mater Des 2014;55:482–91.

[8] Denlinger E, Irwin J, Michaleris P. Thermomechanical modeling of additive manufacturing large parts. J Manuf Sci Eng 2014;136(6):061007.

[9] Denlinger E, Heigel J, Michaleris P, Palmer T. Effect of inter-layer dwell time on distortion and residual stress in additive manufacturing of titanium and nickel alloys. J Mater Process Technol 2015;215:123–31.

[10] Heigel J, Michaleris P, Palmer T. In situ monitoring and characterization of distortionduring laser cladding of Inconel® 625. J Mater Process Technol 2015;220:135–45.

[11] King W, Anderson A, Ferencz R, Hodge N, Kamath C, Khairallah S. Overview of modelling and simulation of metal powder-bed fusion process at Lawrence livermore national laboratory. Mater Sci Technol 2014;31(8):957–68.

[12] Michaleris P. Modeling metal deposition in heat transfer analyses of additive manufacturing processes. Finite Elem Anal Des 2014;86:51–60.

[13] Michaleris P, DeBiccari A. Prediction of welding distortion. Weld J-Includ Weld Res Suppl 1997;76(4):172.

[14] Patil R, Yadava V. Finite element analysis of temperature distribution in single metallic powder layer during metal laser sintering. Int J Mach Tools Manuf 2007;47(7):1069–80.

[15] Peyre P, Aubry P, Fabbro R, Neveu R, Longuet A. Analytical and numerical modelling of the direct metal deposition laser process. J Phys D, Appl Phys 2008;41(2):025403.

[16] Roberts I, Wang C, Esterlein R, Stanford M, Mynors D. A three-dimensional finite element analysis of the temperature field during laser melting of metal powders in additive layer manufacturing. Int J Mach Tools Manuf 2009;49(12):916–23.

[17] Li C, Fu C, Guo Y, Fang F. A multiscale modeling approach for fast prediction of part distortion in selective laser melting. J Mater Process Technol 2016;229:703–12.

[18] Li C, Fu C, Guo Y, Fang F. Fast prediction and validation of part distortion in selective laser melting. Proc Manuf 2015;1:355–65.

[19] Li C, Liu J, Guo Y. Prediction of residual stress and part distortion in selective laser melting. Proc CIRP 2016;45:171–4.

[20] Papadakis L, Loizou A, Risse J, Schrage J. Numerical computation of component shape distortion manufactured by selective laser melting. Proc CIRP 2014;18:90–5.

[21] Neugebauer F, Keller N, Ploshikhin V, Feuerhahn F, Köhler H. Multi scale FEM simulation for distortion calculation in additive manufacturing of hardening stainless steel. In: International workshop on thermal forming and welding distortion; 2014.

[22] Denlinger E, Heigel J, Michaleris P. Residual stress and distortion modeling of electron beam direct manufacturing Ti-6Al-4V. Proc Inst Mech Eng, B J Eng Manuf 2014;229:1803–13.

[23] Dunbar A, Denlinger E, Heigel J, Michaleris P, Guerrier P, Martukanitz R, et al. Development of experimental method for in situ distortion and temperature measurements during the laser powder bed fusion additive manufacturing process. Addit Manuf 2016;12:25–30.

[24] Yilbas B, Akhtar S, Karatas C. Laser surface treatment of Inconel 718 alloy: thermal stress analysis. Opt Lasers Eng 2010;48(7):740–9.

[25] Wang Z, Guan K, Gao M, Li X, Chen X, Zeng X. The microstructure and mechanical properties of deposited-IN718 by selective laser melting. J Alloys Compd 2012;513:518–23.

[26] Pottlacher G, Hosaeus H, Kaschnitz E, Seifter A. Thermophysical properties of solid and liquid Inconel 718 alloy*. Scandinavian J Metall 2002;31(3):161–8.

[27] Garcí V, Arriola I, Gonzalo O, Leunda J, et al. Mechanisms involved in the improvement of Inconel 718 machinability by laser assisted machining (LAM). Int J Mach Tools Manuf 2013;74:19–28.

[28] Ahn D, Byun K, Kang M. Thermal characteristics in the cutting of Inconel 718 superalloy using CW Nd: YAG laser. J Mater Sci Technol 2010;26(4):362–6.

[29] Thomas D. The development of design rules for selective laser melting. University of Wales; 2009.

[30] Jacobs P. Rapid prototyping & manufacturing: fundamentals of stereolithography. Society of Manufacturing Engineers; 1992.

[31] Zhang L, Reutzel E, Michaleris P. Finite element modeling discretization requirements for the laser forming process. Int J Mech Sci 2004;46(4):623–37.

[32] Sainte-Catherine C, Jeandin M, Kechemair D, Ricaud JP, Sabatier L. Study of dynamic absorptivity at 10.6 μm (co2) and 1.06 μm (nd-yag) wavelengths as a function of temperature. J Phys IV 1991;1(C7):C7–151.

15

Validation of the American Makes Builds

Jeff Irwin, Michael Gouge

Product Development Group, Autodesk Inc., State College, PA, United States

15.1 INTRODUCTION

With the emergence of commercial software purporting to predict the distortion of components manufactured using laser powder bed fusion (LPBF) processes, it is necessary to have a public library of parts for modeling tool validation. This was the impetus for the formation of the Engage research conglomerate under America Makes, which is a National Center for Defense Manufacturing and Machining (NCDMM) project [1]. Through the America Makes project, production side research groups manufactured a number of geometries which were then used for validation of LPBF processes by additive manufacturing simulation software companies. This work describes one of the most difficult to model Inconel® 625 geometries from production at the General Electric Global Research Center (GEGRC) and simulated using Netfabb Simulation by Autodesk Inc.

LPBF operates by selectively melting a thin layer of metal powder using a laser [2,3]. After melting and a short cool down period, a new layer of powder is spread out over the build chamber by a recoater blade, which may be rigid, flexible, or even a roller style device depending upon the manufacturer. Then the next layer of metal is melted. The regions of melt are controlled numerically, based upon a CAD geometry, which is converted into thin layers by software which is called a *slicer*. The melting and recoating process repeats until the geometry is completed. This results in a metallic, net or near-net component. However, the finished part may not be exactly as intended as the high thermal gradients induced by melting a small amount of material on a relatively cool part causes thermal stresses which may produce plastic deformation [4–6]. During deposition the hot material expands outward at a much greater rate than the cooler material beneath it, which causes the component and build plate to bend downwards from the center. As the hot material cools again it experiences a much greater contracting thermal strain than the material beneath it, which bends the component and build plate upward. Thermal stresses are higher during the contraction phase than the expansion phase because of the differences in material properties, which implies that the repeated cycles of melting and cooling will deform the part permanently upward. It is common practice to include *support structures* to anchor the part to the build plate to mitigate deformation, particularly for those regions which are not directly supported by material, called *overhang*, which are subject to the greatest amount of deformation [7]. These anchoring builds are sacrificial. Support structures are thus built as a thin, sparse mesh, often with teeth like connections to the part and build plate, so as to ease post-build removal. However, even with a carefully designed geometry and support structures, unwanted plastic deformation often occurs. To further mitigate this, build orientation may be altered, support structures may be added or thickened, or the build component geometry itself may need to be redesigned. This may be very costly as complex builds can easily cost thousands or tens-of-thousands of dollars [8]. Using a validated thermo-mechanical model however may reduce the need for costly experimental iterations.

Finite element (FE) models have been used for decades to mitigate deformation in welding and directed energy deposition (DED) processes [9–14]. The typical methodology is to perform a decoupled thermo-mechanical model. This entails first running a simulation of the thermal history of the deposition process, modeling the heat source, conduction through the part, and balancing surface losses via radiation and conduction. Then this temperature history is used to compute the mechanical response to the thermal loading, i.e. thermal, elastic, and plastic stresses and strains. This method is widely practiced, effective, and accurate. However, when researchers have attempted to apply this modeling technique to LPBF processes it was found to be computationally infeasible for large scale parts. During the development of welding and DED FE models, two heuristics were found to guide modelers. First, the mesh needs to be sized to capture the heat source. Depending upon the modeling software used, the applied rule of thumb is 4–8 elements per laser radius using traditional FE packages, or 1–2 elements using Netfabb Simulation to ensure a converged mesh [15]. Second, to avoid aliasing the time step should be set so that each forward iteration is equal to the time it takes for the laser to run the length of one element [16]. Laser radii for LPBF processes are on the order of 1/10th the size of DED models, so using voxel elements this would require a mesh with the order of 1000 times as many elements as a comparative DED build. Laser deposition rates for LPBF range from 5–100 times DED builds. These two effects would require a tremendous amount of time, processing power, and data storage to complete. This has driven the research to attempt multi-scale modeling methods.

The multi-scale approach rests upon the principle that physical phenomena can be lumped together in time, space, or both time and space. Due to the extremely high laser travel speeds in LPBF, approaching or even surpassing 1000 mm/s, this process has the potential to be modeled using multi-scale methods. A commercially available multi-scale software is used for the present analysis which uses a multi-scale method to model the LPBF process on a layer by layer basis. The modeling process works as follows:

1. A fine-scale moving source thermal analysis of a small volume of material is performed. This models the heat source directly using the above mesh and time step heuristics.
2. A fine-scale moving source mechanical response analysis of the small volume is performed. This model determines the mechanical behavior of the material during laser melting.

3. A part-scale layer by layer thermal analysis of the actual component geometry is performed, using information from the fine-scale thermal model.

4. A part-scale layer by layer mechanical analysis is performed, using modeling information from both the part-scale thermal analysis and the fine-scale mechanical analysis, to determine the distortion of the part during production.

To ensure this multi-sale modeling method is accurate, a validation study must first be performed.

The validation of thermo-mechanical models can be achieved using either in-situ or post-process methods. In-situ measurement techniques, such as using thermocouples or displacement sensors to monitor the development of thermal histories and distortion during manufacturing, allows insights into the physical phenomena during additive manufacturing [17, 18]. Post-process measurements, e.g. using a coordinate measuring machine (CMM), white light scanning or, computerized tomography (CT) scanning, only reveals the distortion of the final product, however these measurements can provide a complete map of distortion as opposed to a single point to which the common in-situ measurement methods are limited. The Netfabb Simulation FE solver has been validated previously using a variety of both in-situ and post-process methods for the modeling of direct energy deposition (DED) and LPBF components [14,19–23].

In-situ validation of additive processes dates to the development of the earliest welding models, as seen in Nickell and Hibbit, who use thermocouples to validate a thermal finite element (FE) welding model [24]. Thermocouples are also effectively applied to the validation of a Tungsten Inert Gas (TIG) welding [25], DED models [26,27]. Infrared video techniques are an alternate technique to capture thermal histories, which is utilized in the validation of multipass welding [28], DED builds [29–31], and LPBF [32] modeling efforts. Validation of the mechanical response for additive manufacturing using in-situ methodologies include application of strain rosettes to the substrate prior to metal deposition [33], or positioning a linear variable differential transformer (LVDT) [34], or laser displacement sensor [31] beneath the substrate. In-situ measurements can yield deeper understanding of the relationship between process parameters and geometry with thermal behavior and the resulting development of residual stresses and plastic deformation. However, these measurement methods can be more difficult to plan and implement and may impose limitations on the geometry to be built.

Post-process validation methods primarily focus upon measuring residual stresses and deformation. Post-process residual stresses can be determined using hole cutting methods [33,35–38], X-ray diffraction [6,13,39,40], or neutron diffraction [29,41]. Residual stresses, while excellent tools for the validation of mechanical models, are difficult to implement as hole drilling methods are limited to certain geometries, and diffraction techniques are often prohibitively expensive to complete. Furthermore many components are stress relieved post-build, limiting the industrial interest in residual stresses. Measurements of deformation however are simple and inexpensive to implement, while providing vital data to the experimentalist about the quality of the build and to the modeler about the quality of the simulation. Optical techniques can be used to compare post-process distortion, such as the photograph–model comparisons in Michaleris and DeBiccari [10], and Alimardani et al. [26]. Direct post-process measurements of deflection can also be effective, however the most comprehensive methods of post-process distortion measurement are 3D scanning techniques, as implemented by Hojny [42]. Three-dimensional scans produce surface maps of finished products which can be compared to the source CAD geometry which was attempted to be built. These scans can be used to investigate part distortion, and can be compared with thermo-mechanical simulation results to validate their predictive capabilities. For these reasons GEGRC has produced scans for the two geometries simulated in the present study using Netfabb Simulation.

Prior to the current work, Netfabb Simulation has been validated for a variety of processes and materials. Inconel® 625 Thermal cladding simulations showed excellent adherence to thermocouple measurements [19,20]. Accurate thermo-mechanical models were shown for thin walls [14], and a very large Ti-6Al-4V EBEAM builds [21]. Dunbar et al. validated a LPBF thermo-mechanical model using in-situ thermocouple and LVDT measure [22], using post-process CMM measurement [23]. However the validation of the LPBF used minimal data and was for a very simple cylindrical geometry.

The objective of this work is to validate the distortion prediction capabilities of Netfabb Simulation using the multi-scale approach and to present a methodology for future validation studies. An overview of the multi-scale FE approach will be presented. Experimental details will be given and the two validation geometries will be shown and discussed. The implementation of the modeling approach will be outlined along with a discourse concerning validation methods and criteria. Finally the experimental and simulation results will be presented, along with comparisons of measured versus modeled distortion, and a discussion of validation methods.

15.2 MODELING APPROACH

The multi-scale model of the LPBF process is performed using Netfabb Simulation, a commercially available software package for the thermo-mechanical FE modeling of AM processes. The moving-source modeling methods detailed in Chapter 2 are used to perform the fine-scale model. These modeling techniques have been extensively validated, as described in Chapters 3–4 and 7, 13–14 of this book. The inputs into the fine-scale model are laser power, laser absorbtion efficiency, laser travel speed, layer thickness, laser beam diameter, gap width, initial rotation angle, interlayer rotation angle, and interlayer recoater dwell time. Thermal boundary conditions are imposed to simulate losses into both the bulk part, the powder, and the ambient atmosphere at the top of the build. Mechanical boundary conditions fix the base of the fine-scale build volume, to simulate the rigidity of the base plate or previously built solid material. After completion, select, proprietary information is passed from the fine-scale model to the part-scale model. The part-scale model input consists of the aforementioned data from the part-scale simulation, build geometry, support structure geometry, and base plate geometry. Multiple layers are grouped together to further reduce computation time. Layers may be grouped, as the fine-scale model reveals inter-layer thermo-mechanical interactions. The thermal analysis has a uniform bulk temperature applied to each layer group, as determined by the fine-scale model. Surface convection and radiation losses are also accounted for, in addition to conduction through the part and into the build plate. Conduction losses from

TABLE 15.1 Processing parameters

Laser power [W]	125
Laser travel speed (mm/s)	500
Laser spot size (mm)	0.09
Gap width (mm)	0.09
Layer thickness (mm)	0.030
Scan strategy	Parallel with 67° interlayer rotation

the build plate into the printer structure itself are approximated as a surface convection. The resulting part-scale thermal analysis and details from the fine-scale mechanical simulation is used to calculate the grouped layer part-scale mechanical prediction of distortion. The base of the build plate is fixed to simulate the bolting of the plate into the machine.

15.3 EXPERIMENTAL PROCEDURES

Validation is performed using data from the America Makes project, "Development of Distortion Prediction and Compensation Methods for Metal Powder-Bed AM". This research consortium, under the heading of the National Center for Defense Manufacturing and Machining (NCDMM), was created with the express view of validating and comparing commercially available AM distortion modeling tools. Several geometries are designed and printed by the GE Global Research Center using the SLM250 ReaLizer, using the processing parameters given in Table 15.1.

15.3.1 Experimental Validation Geometry

The America Makes project determined through both experiments and simulations several geometries that provide good cases for determining the validity of LPBF FE analyses. It was determined that the ideal validation geometry would have both thin and thick sections, exhibit measurable levels of distortion, and be easy to build repeatedly without failure. Of all the test geometries investigated, the ideal candidate found was the Square Canonical part. This part, shown in Figure 15.1, has thin, thick, and transition sections, distorts significantly, and does not crack or experience recoater blade interference problems. This is the geometry that will be used to validate Netfabb Simulation in the present work.

15.3.2 Post-Process Measurement of Distortion

After construction the built components are scanned using a white-light 3D scanner.

Figure 15.2 shows the 3D scan of the Inconel 625 square canonical builds produced by GEGRC.

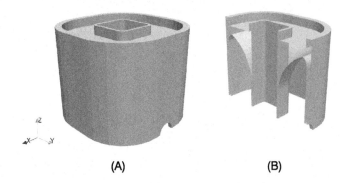

FIGURE 15.1 Validation geometry. (A) Square canonical geometry; (B) Cross-section of square canonical geometry.

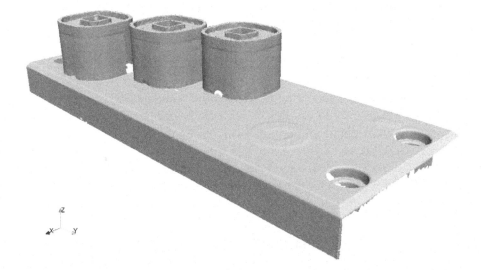

FIGURE 15.2 3D Scan of LPBF square canonical geometry builds for experimental validation.

15.4 NUMERICAL IMPLEMENTATION

Below the relevant details of how the simulation is completed are given. Meshing details and a mesh convergence study of the part scale model is given. A discussion of validation methods to be used for experiment-model comparison is presented.

15.4.1 Automatic Mesh Generation From a Source CAD File

Netfabb Simulation will automatically create a voxel based mesh from a source stereolithography (STL) file. The software allows the end user to control the resulting mesh with two primary inputs: the number of layers to group together and the number of mesh coars-

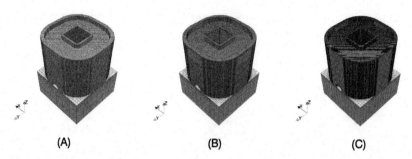

FIGURE 15.3 Autogenerated voxel meshes for the 3 part mesh convergence study. (A) Layers grouped $= 13$; (B) Layers grouped $= 8$; (C) Layers grouped $= 5$.

ening steps. The number of elements grouped together determines the size of the finest elements. This must be sized according to the thinnest feature of the source STL file and the layer thickness during building. For instance, if the thinnest wall of a geometry is 1 mm and the layer thickness is 0.05 mm then at most 20 layers can be grouped together without losing mesh resolution. However this does not guarantee the mesh is adequate to capture the physical behavior. Thus the end user must follow good modeling practice and perform a mesh convergence study for each geometry to be modeled.

For the square canonical geometry the thinnest element is the interior wall. This wall is 0.41 mm. The part was built using 0.03 mm thick layers, which would allow at maximum 13 layers to be grouped together to adequately capture the geometry, rounding down. For the convergence study 3 meshes were looked at grouping 13, 8, and 5 layers together. No coarsening steps were taken so as not to further complicate the convergence study. The resulting meshes are shown in Figure 15.3.

15.4.2 Dwell Time Multiplier

In the experiment, 3 of the square canonical geometries were built. However during simulation, to save computational time and resources, only one of the 3 parts is simulated. To ensure the timing is correct for the simulation, the time between the modeling of the layers is multiplied by 3 to allow for the extra cooling between layer additions.

15.4.3 Validation Criteria

Validation of models typically rely upon both qualitative and quantitative metrics of error. Qualitative error analysis is typically performed by visual trend matching. While this allows for quick and detailed comparisons, trend matching analysis lacks the rigor of numeric comparisons. Quantitative comparisons allow for percent error or other such error analysis methods to be used as a compact datum point for measuring the quality of model results, but are limited by the discrete nature of measurement methods. The present study will use both visual trend matching of distortion plots and quantified percent error analysis of peak distortion.

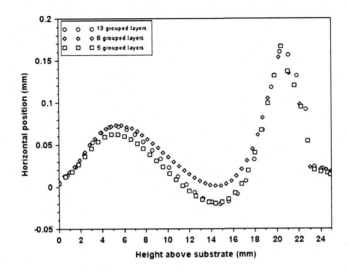

FIGURE 15.4 Mesh convergence study results.

15.5 RESULTS AND DISCUSSION

In this section the results of the mesh convergence study will be given first. Once convergence has been achieved, distortion comparisons between the 3D scans of the experimental build and the simulation will be presented. A discussion of qualitative 2D distortion trend matching will follow, then the peak distortion percent error will be presented. In this way the multi-scale modeling method employed by Netfabb Simulation will be validated.

15.5.1 Mesh Convergence Study Results

The convergence study examines the modeled distortion along the positive X face for the 3 mesh density settings. Results of the mesh convergence study are presented in Figure 15.4.

The convergence study figure shows that the trends are similar and the values seemingly identical for the 3 mesh settings used. The peak distortion values are: 13 grouped layers = 0.160 mm, 8 grouped layers = 0.154 mm, 5 grouped layers = 0.167 mm. From 13 to 8 grouped layers modeled peak distortion, there is a 4.5% difference, which is below the standard metric of convergence, 5%. This shows the mesh has already converged at 13 grouped layers. From 13 to 5 grouped layers there is a 4.3% difference in peak distortion. For the final validation study the finest results will be used to retain the maximum resolution over the height of the build.

15.5.2 Validation Study

Plots of both experimental and simulated distortion along the 4 outer walls are given in Figure 15.5. Along all 4 faces the trends are well correlated between the simulation and ex-

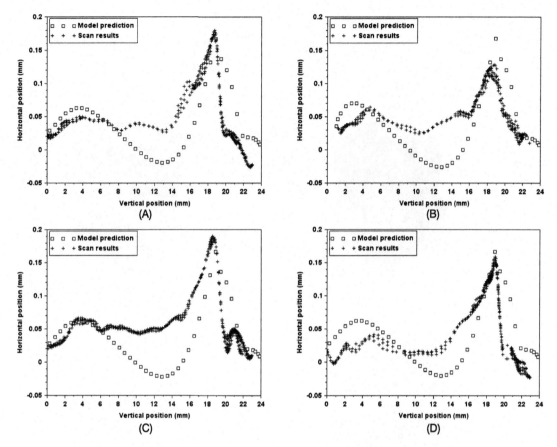

FIGURE 15.5 Square canonical part validation study results. All values are in mm. (A) Displacement results, negative X face; (B) Displacement results, positive X face; (C) Displacement results, negative Y face; (D) Displacement results, positive Y face.

tracted scan data. The distortion plot exhibits 2 distinct peaks, a small peak around 4–6 mm in height, and a rather severe peak around 18–20 mm in height. The first peak is captured well on the negative X face, the positive X face, and the negative Y face. The model overpredicts the experimental scan on the positive Y face. The second peak was well matched on the negative X, negative Y, and positive Y face. This shows good general trend matching. However, all 4 faces exhibit excessive negative distortion around 14–16 mm in build height. This indicates the multi-scale modeling method overpredicts the contraction of the part through the center of the build, particularly at the transition from the thin to thick section. The quantified error analysis at the peak distortion is presented in Table 15.2.

Examining the tabulated quantified error shows that the prediction distortions are within 12% of the measurement at all but 1 of the 4 faces. The worst corresponding location, the negative Y face, does not show the severity of distortion. The lack of model–measurement correspondence may be due to modeling approximations, such as modeling the thermal losses

TABLE 15.2   Quantified validation analysis of the square canonical part

	Negative X face	Positive X face	Negative Y face	Positive Y face
Peak distortion measured (mm)	0.180	0.128	0.189	0.159
Peak distortion modeled (mm)	0.177	0.167	0.166	0.167
Peak distortion% error	−1.4%	30%	−12%	5.2%

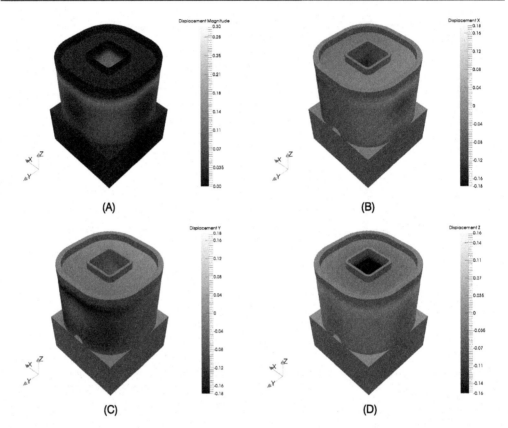

FIGURE 15.6   Final displacement results. (A) Displacement magnitude; (B) Displacement X direction; (C) Displacement Y direction; (D) Displacement Z direction.

into the powder as convection, or build plate location effects. However, there are two faces with errors less than 6%. Taken together with the trend matching, these errors are low enough that the tool may be used for useful distortion predictions. These results may be used to analyze regions of problematic distortion and be used to alter the geometry or build orientation, to investigate if the build quality may be improved before attempting to manufacture the component.

3D plots of distorted simulation results may be seen in Figure 15.6. From the above results there are a couple observations that can be made. First, there is indeed significant distortion of this part. The outer wall is 1.14 mm thick. It displaces, at maximum, 0.189 mm, which is more than 10% of the wall thickness. Second, the distortion in the X and Y directions nearly identical, showing the distortion behavior is roughly symmetric. This indicates that distortion mitigation should be applied uniformly in the X and Y direction.

15.6 CONCLUSIONS

This work has validated the multi-scale modeling approach using Netfabb Simulation. This approach works by first simulating the fine-scale thermal-mechanical behavior on a small amount of representative build volume, then mapping information gleaned from the small volume model to a part-scale model. Modeling was performed using the America Makes Square Canonical geometry, using 3D scans of an Inconel® 625 build provided by GEGRC. Details of the multi-scale modeling method were given. A 3 part mesh convergence study was performed, which determined that a mesh that grouped 5 build layers at a time was required to accurately model the geometry. Overlay comparisons between the 3D scan and the simulation results showed good all around agreement. Trend matching was observed in the 2D plots of displacement along the center of the 4 exterior walls over the build height. A quantitative error analysis comparing peak distortion along the 4 exterior walls indicated the model comes within 12% of the measurement at 3 of the 4 compared locations and within 30% of all of the numerically compared locations. These results indicate that the multi-scale modeling is an accurate simulation tool that can be used to predict and mitigate distortion of Inconel® 625 LPBF parts.

References

[1] America Makes Announces Second Project Call Awardees. Available from, https://www.americamakes.us/news-events/press-releases/item/475-second-projectcall-awardees, 2014.

[2] Kruth JP, Leu MC, Nakagawa T. Progress in additive manufacturing and rapid prototyping. CIRP Ann-Manuf Technol 1998;47(2):525–40.

[3] Levy GN, Schindel R, Kruth JP. Rapid manufacturing and rapid tooling with layer manufacturing (LM) technologies, state of the art and future perspectives. CIRP Ann-Manuf Technol 2003;52(2):589–609.

[4] Shiomi M, Osakada K, Nakamura K, Yamashita T, Abe F. Residual stress within metallic model made by selective laser melting process. CIRP Ann-Manuf Technol 2004;53(1):195–8.

[5] Matsumoto M, Shiomi M, Osakada K, Abe F. Finite element analysis of single layer forming on metallic powder bed in rapid prototyping by selective laser processing. Int J Mach Tools Manuf 2002;42(1):61–7.

[6] Song X, Xie M, Hofmann F, Illston T, Connolley T, Reinhard C, et al. Residual stresses and microstructure in powder bed direct laser deposition (PB DLD) samples. Int J Mater Forming 2015;8(2):245–54.

[7] Gao W, Zhang Y, Ramanujan D, Ramani K, Chen Y, Williams CB, et al. The status, challenges, and future of additive manufacturing in engineering. Comput Aided Des 2015;69:65–89.

[8] Baumers M, Dickens P, Tuck C, Hague R. The cost of additive manufacturing: machine productivity economies of scale and technology-push. Technol Forecast Soc Change 2016;102:193–201.

[9] Ueda Y, Yamakawa T. Analysis of thermal elastic-plastic stress and strain during welding by finite element method. Trans Jpn Weld Soc 1971;2(2):186–96.

[10] Michaleris P, DeBiccari A. Prediction of welding distortion. Weld J-Includ Weld Res Suppl 1997;76(4):172s.

[11] Michaleris P, Dantzig J, Tortorelli D. Minimization of welding residual stress and distortion in large structures. Weld J New York 1999;78:361s.

[12] Lindgren LE. Finite element modeling and simulation of welding. Part 1: increased complexity. J Therm Stresses 2001;24(2):141–92.

[13] Ghosh S, Choi J. Three-dimensional transient finite element analysis for residual stresses in the laser aided direct metal/material deposition process. J Laser Appl 2005;17(3):144–58.

[14] Heigel J, Michaleris P, Reutzel E. Thermo-mechanical model development and validation of directed energy deposition additive manufacturing of Ti-6Al-4V. Addit Manuf 2015;5:9–19.

[15] Goldak J, Chakravarti A, Bibby M. A new finite element model for welding heat sources. Metall Trans B 1984;15(2):299–305.

[16] Michaleris P. Modeling metal deposition in heat transfer analyses of additive manufacturing processes. Finite Elem Anal Des 2014;86:51–60.

[17] Denlinger ER, Heigel JC, Michaleris P. Residual stress and distortion modeling of electron beam direct manufacturing Ti-6Al-4V. Proc Inst Mech Eng, B J Eng Manuf 2015;229(10):1803–13.

[18] Heigel J, Michaleris P, Palmer T. In situ monitoring and characterization of distortion during laser cladding of Inconel® 625. J Mater Process Technol 2015;220:135–45.

[19] Gouge MF, Heigel JC, Michaleris P, Palmer TA. Modeling forced convection in the thermal simulation of laser cladding processes. Int J Adv Manuf Technol 2015;79(1–4):307–20.

[20] Gouge M, Michaleris P, Palmer T. Fixturing effects in the thermal modeling of laser cladding. J Manuf Sci Eng 2017;139(1):011001.

[21] Denlinger ER, Irwin J, Michaleris P. Thermomechanical modeling of additive manufacturing large parts. J Manuf Sci Eng 2014;136(6):061007.

[22] Dunbar A, Denlinger E, Heigel J, Michaleris P, Guerrier P, Martukanitz R, et al. Development of experimental method for in situ distortion and temperature measurements during the laser powder bed fusion additive manufacturing process. Addit Manuf 2016;12:25–30.

[23] Dunbar AJ, Denlinger ER, Gouge MF, Michaleris P. Experimental validation of finite element modeling for laser powder bed fusion deformation. Addit Manuf 2016;12:108–20.

[24] Nickell RE, Hibbitt HD. Thermal and mechanical analysis of welded structures. Nucl Eng Des 1975;32(1):110–20.

[25] Anca A, Fachinotti V, Escobar-Palafox G, Cardona A. Computational modelling of shaped metal deposition. Int J Numer Methods Eng 2011;85(1):84–106.

[26] Alimardani M, Toyserkani E, Huissoon JP. A 3D dynamic numerical approach for temperature and thermal stress distributions in multilayer laser solid freeform fabrication process. Opt Lasers Eng 2007;45(12):1115–30.

[27] Peyre P, Aubry P, Fabbro R, Neveu R, Longuet A. Analytical and numerical modelling of the direct metal deposition laser process. J Phys D, Appl Phys 2008;41(2):025403.

[28] Bai X, Zhang H, Wang G. Improving prediction accuracy of thermal analysis for weld-based additive manufacturing by calibrating input parameters using IR imaging. Int J Adv Manuf Technol 2013;69(5–8):1087–95.

[29] Wang L, Felicelli SD, Pratt P. Residual stresses in LENS-deposited AISI 410 stainless steel plates. Mater Sci Eng A 2008;496(1):234–41.

[30] Hammell JJ, Ludvigson CJ, Langerman MA, Sears JW. Thermal imaging of laser powder deposition for process diagnostics. In: ASME 2011 international mechanical engineering congress and exposition. American Society of Mechanical Engineers; 2011. p. 41–8.

[31] Lundbäck A, Lindgren LE. Modelling of metal deposition. Finite Elem Anal Des 2011;47(10):1169–77.

[32] Roberts I, Wang C, Esterlein R, Stanford M, Mynors D. A three-dimensional finite element analysis of the temperature field during laser melting of metal powders in additive layer manufacturing. Int J Mach Tools Manuf 2009;49(12):916–23.

[33] Grum J, Šturm R. A new experimental technique for measuring strain and residual stresses during a laser remelting process. J Mater Process Technol 2004;147(3):351–8.

[34] Plati A, Tan J, Golosnoy I, Persoons R, Van Acker K, Clyne T. Residual stress generation during laser cladding of steel with a particulate metal matrix composite. Adv Eng Mater 2006;8(7):619–24.

[35] Rybicki E, Shadley J. A three-dimensional finite element evaluation of a destructive experimental method for determining through-thickness residual stresses in girth welded pipes. J Eng Mater Technol (United States) 1986;108(2).

[36] Pinkerton A, Li L. An analytical model of energy distribution in laser direct metal deposition. Proc Inst Mech Eng, B J Eng Manuf 2004;218(4):363–74.

[37] Elcoate C, Dennis R, Bouchard P, Smith M. Three dimensional multi-pass repair weld simulations. Int J Press Vessels Piping 2005;82(4):244–57.

[38] Chiumenti M, Cervera M, Salmi A, Agelet de Saracibar C, Dialami N, Matsui K. Finite element modeling of multi-pass welding and shaped metal deposition processes. Comput Methods Appl Mech Eng 2010;199(37):2343–59.

[39] Labudovic M, Hu D, Kovacevic R. A three dimensional model for direct laser metal powder deposition and rapid prototyping. J Mater Sci 2003;38(1):35–49.

[40] Foroozmehr E, Kovacevic R. Effect of path planning on the laser powder deposition process: thermal and structural evaluation. Int J Adv Manuf Technol 2010;51(5–8):659–69.

[41] Ding J, Colegrove P, Mehnen J, Ganguly S, Almeida PS, Wang F, et al. Thermo-mechanical analysis of wire and arc additive layer manufacturing process on large multi-layer parts. Comput Mater Sci 2011;50(12):3315–22.

[42] Hojny M. Thermo-mechanical model of a TIG welding process for the aircraft industry. Archives Metall Mater 2013;58(4).

Appendix

A.1 MATERIAL PROPERTIES USED IN FE MODELING

A.1.1 Aluminum 6061 Properties

TABLE A.1 Thermal properties of Aluminum 6061 [1]

T (°C)	k (W/m °C)	C_p (J/kg °C)
0	162	917
98	177	978
201	192	1028
316	207	1078
428	223	1133
571	253	1230

A.1.2 Inconel® 625 Properties

TABLE A.2 Thermal properties of Inconel® 625 [2,3]

T (°C)	k (W/m °C)	C_p (J/kg)
−18	9.2	402
21	9.8	410
38	10.1	414
93	10.8	427
204	12.5	456
316	14.1	480
427	15.7	496
538	17.5	513
649	19.0	560
760	20.8	590
871	22.8	620
982	25.2	645
Density (kg/m³)	8440	
Emissivity	0.42	
Latent heat of fusion (J/kg)	302000	

Thermo-Mechanical Modeling of Additive Manufacturing
DOI: 10.1016/B978-0-12-811820-7.00026-4

TABLE A.3 Mechanical properties of Inconel® 625 [2]

T (°C)	CTE (mm/mm/°C)
20	Reference temperature
20	0.128E–04
93	0.128E–04
204	0.131E–04
316	0.133E–04
427	0.137E–04
538	0.140E–04
649	0.148E–04
760	0.153E–04
871	0.158E–04
927	0.162E–04

T (°C)	Elastic modulus (GPa)	Poisson's ratio
21	208	0.278
93	204	0.280
204	198	0.286
316	192	0.290
427	186	0.295
538	179	0.305
649	170	0.321
760	161	0.340
900	116	0.340
1000	87	0.340
1200	20	0.340

T (°C)	Yield strength (MPa)	Yield strain (mm/mm)	UT Strength (MPa)	UT strain (mm/mm)
0	483	0.002	965	0.48
100	469	0.002	937	0.47
200	428	0.002	896	0.46
300	414	0.002	896	0.54
400	414	0.002	903	0.45
500	414	0.002	910	0.45
600	410	0.002	897	0.50
700	421	0.002	593	0.36
800	410	0.002	414	0.48
871	276	0.002	276	0.01
931	138	0.002	138	0.01
1093	83	0.002	83	0.01

A.1.3 Inconel® 718 Properties

TABLE A.4   Thermal properties of Inconel® 718 [3,4]

T (°C)	k (W/m °C)	C_p (J/kg)
20	11.4	427
100	12.5	441
300	14.0	481
500	15.5	521
700	21.5	561
Density (kg/m³)	8146	
Emissivity	0.58	
Latent heat of fusion (J/kg)	227000	

TABLE A.5   Mechanical properties of Inconel® 718 [4]

T (°C)	CTE (mm/mm/°C)
26	Reference temperature
20	0.128E–04
93	0.128E–04
204	0.135E–04
316	0.139E–04
427	0.142E–04
538	0.144E–04
649	0.151E–04
760	0.161E–04
927	0.162E–04

T (°C)	Elastic modulus (GPa)	Poisson's ratio
21	208	0.278
93	204	0.280
204	202	0.286
316	194	0.290
427	186	0.295
538	179	0.305
649	172	0.321
760	162	0.340
871	127	0.340
954	178	0.340
1200	20	0.340

(continued on next page)

TABLE A.5   (*continued*)

T (°C)	Yield strength (MPa)	Yield strain (mm/mm)	Yield strength 2 (MPa)	Yield strain 2 (mm/mm)
94	1172	0.002	1378	0.20
427	1089	0.002	1323	0.20
537	1068	0.002	1276	0.20
649	1034	0.002	1172	0.20
760	827	0.002	847	0.20
871	276	0.002	286	0.20
982	138	0.002	148	0.20
1093	69	0.002	69	0.20

A.1.4 SAE 304

TABLE A.6   Thermal properties of SAE 304 [5]

Thermal conductivity (k (W/m °C))	16.0
Density (kg/m^3)	8000
Specific heat capacity (J/kg/°C)	500
Elastic Modulus (GPa)	193
CTE Reference T °C	20
CTE (mm/mm/°C)	1.87E−5

A.1.5 Ti-6Al-4V

TABLE A.7   Thermal properties of Ti-6Al-4V [6]

T (°C)	k (W/m °C)	C_p (J/kg)
20	6.6	565
93	7.3	565
205	9.1	674
315	10.6	603
425	12.6	649
540	14.6	699
650	17.5	770
Density (kg/m^3)	4430	
Emissivity	0.54	
Latent heat of fusion (J/kg)	365000	
Stress relaxation temperature °C	690	

TABLE A.8 Mechanical properties of Ti-6Al-4V[6]

T (°C)	CTE (mm/mm/°C)
30.5	Reference temperature
20	0.096E–04
500	0.097E–04

T (°C)	Elastic modulus (GPa)	Poisson's ratio
0	105	0.340
800	62.8	0.340

T (°C)	Yield strength (MPa)	Yield strain (mm/mm)	UT Strength (MPa)	UT strain (mm/mm)
0	770	0.002	770	0.01
800	417	0.002	417	0.01

References

[1] Matweb. Constellium alplan 6061 rolled precision aluminum plate, milled both sides. Technical report, 2014.

[2] Special Metals. Inconel alloy 625. Technical report publication number SMC-063, 2006.

[3] Omega Engineering Inc. Non-contact temperature measurement, vol. 1. Technical report transactions Vol 1, 2nd edn. 1998.

[4] Special Metals. Inconel alloy 718. Publication No. SMC-045, Huntington, WV, USA: Special Metals Corporation; 2007.

[5] AKsteel. Product data sheet 304/304l stainless steel uns s30400/uns s30403. Technical report, 2007.

[6] Welsch Gerhard, Boyer Rodney, Collings EW. Materials properties handbook: titanium alloys. ASM international; 1993.

Index